D0935831

THE LIONS OF TSAVO

THE LIONS OF TSAVO

Exploring the Legacy of Africa's Notorious Man-eaters

Bruce D. Patterson

McGraw-Hill

New York Chicago San Francisco
Lisbon London Madrid Mexico City Milan
New Delhi San Juan Seoul Singapore
Sydney Toronto

The McGraw-Hill Companies

1 2 3 4 5 6 7 8 9 0 AGM/AGM 0 9 8 7 6 5 4

ISBN 0-07-136333-5

Quebecor was printer and binder.

This book is printed on acid-free paper.

To my parents,
who gave me the freedom and
means to become a naturalist,
and to all of my family,
for tolerating all the field seasons since

Contents

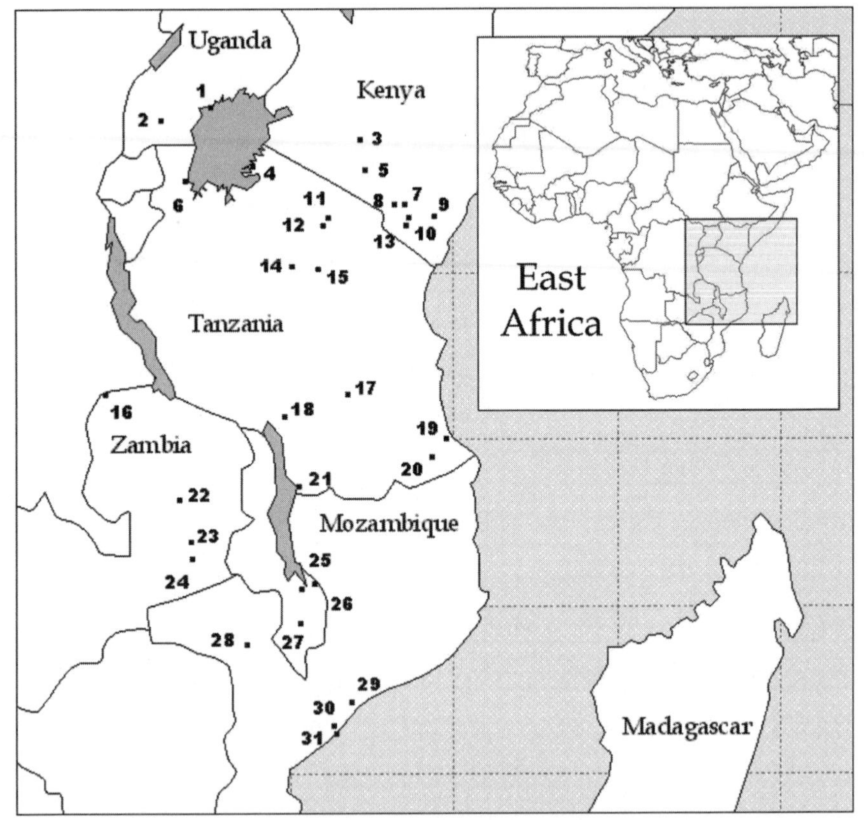

Some of the localities in East Africa where lions have claimed human victims. The plot is by no means exhaustive, and at some localities (7, Tsavo; 18, Njombe), hundreds of deaths were recorded .

FOREWORD

In 1925, the Field Museum received a wonderful gift of two worn lion skins sent by their owner, Colonel John H. Patterson. Colonel Patterson needed to recoup financial loss—the Great Depression was looming after the long and senseless European War of 1914–18. To the receiving institution, securing the skins immortalized by the donor's book, *The Man-eaters of Tsavo,* was an incredible opportunity. The year 1998 marked the beginning of the second century since Colonel Patterson successfully hunted and killed the two most famous man-eating lions of Kenya, known in Patterson's time as British East Africa. That same year also saw the commencement of a collaborative research program between The Field Museum, the Kenya Wildlife Services, and the National Museums of Kenya that would produce the book you are now holding. *Lions of Tsavo: Exploring the Legacy of Africa's Notorious Man-eaters* embodies both tradition and promise: a tradition of research excellence and dissemination of the fruits of such labor to both the scientific and public audiences.

The work of Dr. Bruce D. Patterson represents a milestone in publication, for it is among the first Field Museum scientific books that are designed for both popular and scholarly audiences. It is also a significant contribution to the Field Museum's Africa portfolio. The Field Museum sent its first expedition in 1896 to Africa's British Somaliland, only three years after the Museum's founding. The expedition opened the doors to the long and fruitful engagement of Africa by one of the world's a leading Natural History Museums. That tradition has continued almost unbroken to the present day. The Field Museum presently boasts one of the largest and perhaps most intensive research programs in Africa, and the institution's zoological and anthropological collections from that continent easily rank among the world's finest in many areas.

From the scientific perspective, *The Lions of Tsavo* serves three important objectives. First, it places natural history museums at the forefront of zoological innovation. Second, it bridges the academic divide between ecology, conservation biology, and genetics. Finally, it documents and disseminates knowledge about the world's most gregarious cats, *Panthera leo*, in lucid prose. This volume is an excellent addition to the ever growing list of books on Africa's most storied animals by an outstanding scientist who has been bitten by the Africa bug and fallen head over heels for this enchanted land.

In a century when biologists are taking advantage of advances in technology to secure previously unobtainable data from material objects now preserved only in museum collections, in a world where research funding is now more than ever driven by the bottom-line and research initiatives shaped by political ideology; in an academic environment where basic biology has become antiquated in favor of molecular biology and biotechnology, it is comforting to find a book that discusses basic biology but also competently integrates conservation and molecular biology to address lion behavior and ecology. From this standpoint, *The Lions of Tsavo* represents an outstandingly successful example of interdisciplinary and cross-disciplinary research.

The book fulfills the important goal of addressing ways of preserving biological and cultural diversity currently threatened by extinction (Leakey 2001; Western 1997). Conservation biology and cultural diversity require a scientific understanding of the interdependent relations among biological species and the role of the environment on intra- and inter-species interactions. Dr. Patterson's book contributes to this topic by analyzing lion behavior and ecology in order to scientifically "explain the distinctive appearance and behavior of Tsavo lions." His concern for the future survival of the lions in Africa obligated him to design a long-term project in order to collect as much information as possible on the daily life of Tsavo lions. As such, he and his collaborators and Earthwatch volunteers have spent countless hours following several lion prides and collecting data on intra- and inter-pride interactions and conflict and their hunting, feeding, and mating habits. Because lions in Kenya live in close proximity to people, our awareness of interspecies conflict is now more than ever a matter of public concern and a priority among international conservation agencies.

The Lions of Tsavo is a scientific exploration—from the perspectives of ecology, conservation biology, and molecular biology—of the fabulous lions immortalized by Colonel John H. Patterson. The two lions, whose skulls and skins are curated by the author at The Field Museum, are believed to have had an unusual appetite for people. According to Colonel Patterson, the two male lions wreaked havoc by unleashing a "Reign of Terror"—hunting, killing, and eating railway workers. It took the Colonel nine long and arduous months to kill these lions. The lion story is steeped in myths, superstition, and symbolism. And it is a story of the British colonization of East Africa and the relationships that developed during the process of colonization. It is a story of how Europe tamed and was in turn humbled by Africa.

Nowadays, colonization is a dirty word. To cite the late Elspeth Josceline Grant Huxley:

> *Colonialism is now a dirty word to many, arousing feelings of indignation in black breasts and guilt in white ones—emotions equally disruptive, in my opinion, to a calm assessment of past history and profitable conduct of present affairs. The most cogent summing up of colonialism I have seen was handed down by the quarterchief of Wum in Cameroon to the indefatigable traveller Devla Murphy in the words: "Colonialism is like a zebra. Some say is it a black animal, some say it is a white animal, and those whose sight is good, they say it is a striped animal" (Huxley 1990:xxv; Nichols 2002:xiii).*

I grew up in independent Kenya and did not witness first-hand the might of the British Empire and its assumed civilizing mission. However, over the years I have read and studied enough to know that major historical and ecological transformations have occurred during the past 150 years. Because of its relatively healthy climate, fertile soils, and proximity to Europe, Kenya was one of the most favored colonies and attracted a wealthy European settler community. By the 1920s, the European community was numerous enough to request that the British Secretary of Colonies declare unilateral independence and self-rule from Britain. Led by Hugh Cholmondeley, fondly remembered by some as Lord Delamere, and other wealthy landowners, the Europeans had appropriated the most fertile areas of the country for the exclusive farming of Europeans.

European fascination with Africa can be traced back to the Victorian period when the myth of a wild Africa teeming with riches and savagery was firmly imprinted (Adams and McShane 1996). For Kenya, Dr. Margery Perham (1976:190) demystifies wild Africa in her sober account written on September 29, 1930 while on a train in Kenya.

> *I am once again amazed by Kenya. We are just crossing the equator yet I look out on what might well be Surrey or Dorset, or perhaps from the crimson soil, Devon. Gentle hills, light woodland, thickly turfted meadows, little streams sparkling over rocks, tall weeds flowering yellow and purple, mossy boulders. It seems to be of the very texture of Western Europe, lying drenched by the night's rain under glorious tropical morning sunlight. It makes me almost feel I must become a settler. In all my travels there is no part of the world where I have felt like this....*
>
> *I think now I understand the 'immigrant community.' To own a bit of this lovely, virgin country; to make a house, and, still more, a garden in which you can mingle all the beauties of Western and tropical flowers; to have a part share in this thrilling sunlight; to have cheap, apparently reverential, impersonal labour; to feel the sense of singularity, of enhanced personality that comes from having a white skin among dark millions. You leave behind the fogs, the conventions, the problems of the old country: your resulting sense of freedom and joy is increased by the space and the sun....And—almost best of all—just round the corner is the wild Africa that not even the commercialization of big-game hunting for American tourists has spoiled, the majesty of the wilds, their strangeness, loneliness and danger. Here, in the effort to kill or to see animals, you spend long days, matching your endurance, skill and courage against theirs, with evenings round the camp fire, and nights in a tent into which, as you turn over on your camp-bed, comes the occasional music of the animals' own night-hunting in the bush.*

In order to pay the huge expense of running the colony, Kenya's first British Commissioner, Mr. Charles Eliot, appealed to wealthy Europeans to emigrate to Kenya. He also introduced nature safaris and commercial hunting to supplement the budget provided by the Secretary of Colonies Office (Steinhart 1989: 252–255). He succeeded in both ventures. Before the First World War, hunting was primarily focused on exploitation of big game. During the early decades of the twentieth century and the late nineteenth century, adventurers, bounty hunters, traders and administrators hunted wildlife for subsistence, trophies and ivory. After two decades, it became clear that unregulated big game hunting threatened the survival of wildlife in Kenya. The first game reserves were established in Kenya and Tanganyika in 1896, making hunting without a license illegal. The development of colonial settlements led to the clearing of large tracts of land to create farmland. This destroyed the habitats of many animals and initiated large-scale extinctions. European farmers killed wildlife with impunity, arguing that the wild animals threatened their investments by killing livestock and spreading diseases. They lobbied for the passing of laws to protect their interests while simultaneously disenfranchising indigenous peoples through unfair laws.

The Game Department was established in the first decades of the 1900s to regulate hunting through licensing and control of ivory collection and export. For many years, half of the department's revenue came from licensing of big game hunting. Many game rangers were skilled hunters, trackers, and sportsmen (Beard 1988; Steinhart 1989). By the 1930s, the alienation of most productive lands in Kenya had been almost completed, boundaries of native reserves had been established, and most of the wildlife in settler lands had almost been hunted to extinction. Concern for loss of habitat and biodiversity gained local and international support when it became clear that the majestic African wildlife might soon disappear.

The Royal National Parks were created in Kenya after the Second World War, and Mervyn Cowie was appointed the first director. In 1948, Tsavo was gazetted as a park, becoming the second Royal National Park in Kenya. The establishment of Tsavo National Park restricted the access of local peoples—the Akamba, Taita, and Waata—to the park, leading to economic decline. People evicted from areas now designated as the park were subjected to famine, poverty, and psychological stresses (Eriksen et al 1996: 212). Land alienation caused changes in diet, economies, social relations, and recreation and broke many of the bonds that people had with the natural and supernatural world (Adam and McShane 1996:31–32 ; Lee and Hitchcock 1998). This brief historical background provides the context in which to evaluate *Lions of Tsavo*, especially since the tensions between people and wildlife threaten the latter's survival in Tsavo and Kenya and in other regions.

Linking Man-eating Behavior to Ecological Crises

The relationship between man-eating behavior among big cats and ecological crisis was first raised by Colonel James Corbett in the *Maneaters of Kumaon* (1946). The book discussed the two man-eating leopards of Kumaon, which between them killed five hundred and twenty-five human beings. Colonel Corbett noted that leopards are "quite capable of carrying their human victims for long distances." Leopards, unlike tigers, are, to a certain extent, scavengers and become man-eaters by acquiring a taste for human flesh when unrestricted slaughter of game has deprived them of their natural food. Corbett then hypothesized conditions under which the habit of man-eating may have developed among Indian leopards. Among the Hindu peoples, generally the dead are cremated. However, in circumstances caused by shortage of fuel or labor or during epidemics when inhabitants

> *die faster than they can be disposed of, a very simple rite, which consist of placing a live coal in the mouth of the deceased, is performed in the village and the body is then carried to the edge of the hill and cast into the valley below. A leopard, in an area where his natural food is scarce, finding these bodies very soon acquires a taste for human flesh, and when the disease dies down and normal conditions are established, he very naturally, on finding his food supply cut off, takes to killing human beings. Of the two man-eating leopards of Kumaon, which between them killed five hundred and twenty-five human beings, one followed on the heels of a very severe outbreak of cholera, while the other followed the mysterious disease which swept through India in 1918 and was called 'war fever.'(p.xix)*

Colonel Corbett's description mirrors several events that occurred in the Tsavo region in the years preceding the building of the railway, namely: drought and famine, epidemics including rinderpest and smallpox, and caravan trade through Tsavo.

The Tsavo region suffered severe drought and famine in the 1860s, 1870s, and into the 1890s. During the 1860s drought, cholera and plague brought by Swahili caravans ravaged the region, affecting ethnic groups such as the Maasai, who lived further inland. The Taita recall the droughts of the 1880s and 1890s as the worst droughts ever. C.W. Hobley, who visited the area in 1892, six or seven years after the drought, reported on the resulting famine, called the Mwakisenge Famine. Large-scale emigration, massive mortality, and sale of children into slavery for food are recalled in the oral traditions. One informant said that there were people dying everywhere: in houses, on roadsides, in gardens, etc. There were dead bodies everywhere. Those who died were left unburied, as no one had the strength to dig graves. The number of bodies was too numerous for hyenas alone to dispose of. The air was said to be filled with the smell of decomposing bodies (Hobley 1895; Merritt 1975).

People began killing each other in competition for food. Many Taita people emigrated for relief to Ukambani, Taveta, Chagga, Pare, and Giriama, where the vast majority claim to have originated. Their homes reverted to wilderness. Those few who remained in and around the area were often attacked by the Kamba, Swahili, and other opportunists, and sold into slavery. Historian Hollis Merritt reports that the population of Taita was about 10,000 before the famine and about 1,000 afterwards.

The situation returned to normal until the 1897–1900 drought and famine, recalled by survivor Thomas Mawora of Kivingoni village in Mwanda as the mother of all famines. The rains failed for two successive summers, ravaging large parts of Ukamba province; extending from Jubaland in the north into German territories to the south, and from the coast inland towards Kikuyuland. Called by several names: "*Njala ya Madumbu*," "*Njala ya Kibaba*," "*Njala ya Mbelele*," the famine brought out an "every man for himself" mentality among the people never before witnessed by the missionaries and government administrators. People killed out of desperation. Parents killed their own children because they could not feed them. Cannibalism was reported. Groups of men would wait in the bush for young men to come near, then would attack them and cut-off their heads, hands and feet so no one could identify the victim if the group was caught. Jiggers, smallpox, and venereal diseases, whose spread was attributed to the railway, erupted early on in the famine. Those who died were not buried but were placed in a cave and covered with stone, the fear being that, if buried, the smallpox would not go away.

In *Lions of Tsavo*, Dr. Patterson tests a number of hypotheses, including Colonel Corbett's hypothesis linking man-eating behavior to ecological crises and cultural stability, and the role of injury, disease, pathology, and old age on lion feeding behaviors. Although the verdict is still out, it is very probable that Tsavo's man-eating lions got their appetite from eating victims of famine, warfare, and the caravan trade. Abrupt climate change in East Africa caused severe ecological changes which affected both man and beast and helped shape the events that occurred in 1898. The colonel and his men came to East Africa at the height of all these changes and paid the price that many native peoples had before them.

Lions of Tsavo represents Dr. Patterson's continuing focus on the recording and preserving of endangered biological phenomena. Although funding for basic research in ecology and behavior has declined over the years, *Lions of Tsavo* reminds us that shirking this important responsibility for sexy molecular biotechnological modeling might not be a good idea. There is a real possibility that we may bear witness to the extinction of biological diversity at the time when we possess the knowledge and resources to forestall, at least for the time being, continued extinction. As you read this book and peruse the pictures, you may revel in the beauty and complexity of the daily life of lions as they navigate an increasingly hostile world. I hope you will be moved to appreciate the importance of the preservation and conservation of biological and cultural diversity.

Lions of Tsavo not only highlights the lions' ecology and behavior, including man eating, but also provides a basis from which a rapprochement between conservationists and indigenous peoples might begin. The future of the conservation of the region's biological and cultural diversity depends on policies and strategies incorporating the needs of peoples and wildlife. The predicament faced by wildlife management officials in Tsavo National Park highlights the most important crisis that we face in the twenty-first century: preserving cultural and biological diversity. How can people and wildlife coexist and share the same limited resources? More specifically, is it possible to maintain wildlife solely inside the park, and people solely outside? Can people and animals coexist if one or the other penetrates the boundaries? Can conservation of biodiversity occur without community support and participation? *Lions of Tsavo* provides an excellent discussion of all these pressing issues.

It seems eminently clear that long-term solutions to the human–wildlife crisis may require a decolonization of perspectives on conservation and the patronizing attitude conservators have for indigenous peoples. It is unnecessary to present the predicament of wildlife as one that can be solved by a few people. The predicament of Africa's wildlife cannot be solved by creating orphanages for juvenile wildlife. Rather, it lies in stopping the killing of adult wildlife. It is a collective problem that requires the cooperation of all peoples for a solution. The solution should be found in the participation of all communities concerned.

Human suffering and poverty in the Taita-Taveta district, where *Lions of Tsavo* is set, is very real. Rapid population growth has shortened and, in many cases, made impossible the use of fallow periods in areas to which the people were relocated when the parks were created. Deforestation, resulting from charcoal burning and severe soil erosion, has depleted soil fertility. People have adjusted to overpopulation by moving to new areas with poor soils and erratic rainfall. Every year they prepare their farms but cannot harvest. Wildlife feasts on their crops just as they are beginning to prepare their granaries. The hot sun usually claims the rest. In the event that wildlife or the scorching sun does not take them, the harvests are so small that they cannot sustain a family for a whole year. The nutrient-depleted land cannot be sub-divided among large extended families forever. To compound the problem, since the present generation of young men and women in Taita-Taveta district is comparatively less educated than previous generations, they have few avenues open to them for overcoming their poverty. Preserving the cultural and biological diversity that has made Kenya famous for culturalists and naturalists is at stake. *The Lions of Tsavo* contributes to this debate. Its accessible and unpretentious prose places it among the best books on lion ecology and behavior. In Dr. Bruce Patterson, the maneless lions of Tsavo have at last found a fitting biographer.

Chapurukha M. Kusimba
Oak Park, Illinois
December 1, 2003

Notes

Adams, J.S and T.O. McShane. 1996. *The Myth of Wild Africa: Conservation without Illusion*. University of California Press, Berkeley.

Beard, P. 1988. *The End of the Game: The Last Word From Paradise*. Thames and Hudson.

Corbett, James. 1946. *The Maneaters of Kumaon*. Oxford University Press, London.

Eriksen, S., E. Ouko, and N Marekia. 1996. Land Tenure and Wildlife Management. In C. Juma and J.B. Ojwang (eds). *In Land We Trust: Environment, Private Property, and Constitutional Change*. Nairobi: Initiative Publishers. pp. 199-230.

Hobley, C. W. 1895. Upon a Visit to Tsavo and the Taita Highland. *Geographical Journal* 5:545-561.

Huxley, Elspeth. 1935. *Whiteman's Country: Lord Delamere and the Making of Kenya. Macmillan, London (2 vols)*

Huxley, Elspeth. 1990. *Nine Faces of Kenya* (Anthology). Collins Harvill, London.

Leakey, R. E. and V. Morell. 2001. *Wildlife Wars: My Fight to Save Arica's Natural Treasures*. St. Martin's Press, New York.

Lee and Hitchcock 1998. In Connah, G.1998. *Transformations in Africa: Essays on Africa's Later Past*. Leicester University Press, Leicester. pp. 14-45.

Merritt, H. E. 1975. *A History of the Taita of Kenya to 1900*. Ph.D. Thesis. Department of History. Bloomington: Indiana University Press.

Nichols, C.S. 2002. *Elspeth Huxley; A Biography*. St. Martins Press, New York.

Perham, M. 1986. *East Africa Journey*. Faber and Faber, London.

Steinhart, E.I. 1989. Hunters, Poachers and Gamekeepers: Towards a Social History of Hunting in Colonial Kenya. *Journal of African History* 30: 247-264.

Western, D. 1997. *In the Dust of Kilimanjaro*. Island Press, Washington, D.C.

Introduction: A Confluence at Tsavo

We heard the vehicle some minutes before we first saw it, its laboring engine, thundering body panels, and pounding tires signaling its approach on the dirt track below. Alex Mwazo and I had been preparing for the arrival all morning—organizing equipment and supplies, collating operating manuals and instructions, printing data sheets, and reviewing tent assignments with Sammy, Galla Camp's manager. Shortly, a plume of dust appeared in the northern sky, creating a faint violet stain on the brilliant, uniformly blue sky. It had not rained for weeks, and what humidity we received each evening from the Indian Ocean, a hundred miles to the east, had long since burned off Africa's sun-drenched plains.

The Land Rover's engine was deafening now as it reached the foot of the *kopje*, the seventy-foot-tall rock outcrop on which our camp was perched, and began to climb it. But at two o'clock in the afternoon, the noise broke only the din of grasshoppers and cicadas, stridulating now from the shade of trees. Even lizards seek shelter from the equatorial sun, and the camp's staff were all off making their own preparations for the arrivals. Jutting high above the surrounding woodlands, the *kopje* always caught a breeze, moderating even the oppressive heat of midday. It was a perfect location for a permanent tented camp, and an exceptional staff tended ours, even by Kenya's lofty standards. All were employees of Savannah Lodges and Camps, one of Kenya's most reputable and environmentally responsible firms. The camp had huge canvas tents with separate thatch roofs to keep them cool, real beds with mattresses, solar-powered lights, and adjoining private bathrooms with flush toilets and hot and cold running water. For a well-traveled field biologist, their opulence and comfort was almost embarrassing, but I knew I could adapt. Actually, the camp's greatest luxury for most guests was the propane-powered refrigerator in the dining tent, which permitted cold drinks at the end of a long dusty day.

"Ready or not, here they come," I said to Alex. We both were in positions to be nervous just then. In the vehicle were seven American and British volunteers, coming to East Africa to assist us in studying lions. I'd met Alex the previous year, when I gave a lecture on my lion work at the Education Centre of Tsavo East National Park, Kenya's largest national park, 12 miles north of our *kopje.* Alex had just graduated from Kenyatta University's ecology program in Nairobi (Kenya's capital city) and sought related employment back home in Voi, where Tsavo's headquarters are located. For the better part of a year, he had been a volunteer at the Tsavo Research Centre, supervised by Dr. Samuel Kasiki, the Centre's director and a collaborator on the lion studies. When our project was funded, Kasiki selected Alex to be our graduate assistant. His responsibilities included scientific guidance, training, and supervision of the arriving volunteers.

If this was Alex's bread and butter, it was only slightly less important for me. As the principal investigator of the "Lions of Tsavo" project, I'd spent the better part of two years assembling the team and resources needed to study the ecology and behavior of Tsavo's remarkable lions. With Roland Kays of the New York State Museum and Samuel Kasiki of the Kenya Wildlife Service (KWS), I had proposed a research plan that received the support of Earthwatch Institute* earlier in the year. Earthwatch coordinates volunteers from around the world to assist in scientific fieldwork. Each year, it places about thirty-five hundred volunteers in 120 different projects in fifty countries worldwide, providing them with appropriate briefings, travel coordination, and the like. Although the volunteers typically lack technical backgrounds or expertise, they provide energy, enthusiasm, and, by paying their own expenses plus a bit more, operating funds for projects like ours.

That was the theory, anyhow, but you couldn't prove it by me, because I'd never run an Earthwatch project. And I wasn't just worried about ensuring that the data were collected properly either. In two decades of fieldwork in remote places, such as western Amazonia and islands south of the Magellanic Straits, I had seen field teams compromised by shirkers, malingerers, and whiners, and all of *those* people had credentials and were drawn from bona fide scientific ranks! How selective was Earthwatch in choosing the volunteers? The application materials they submit could only tell you so much. Had these people ever spent time in the bush? Shaken out their

*www.earthwatch.org

shoes to dislodge sheltering scorpions? Taken care not to step on snakes or walk into thorns? Only that morning, about three o'clock, we had heard a lion roaring at the foot of the *kopje*—how many of these volunteers would sleep restfully in their tents with all the sounds of Africa? And with the diverse and unpredictable demands of establishing a new project, what sort of a teacher would I be? Would I find the patience to coach fumbling fingers? The arriving vehicle was bringing the first of *nine teams* of volunteers that would join us in 2002 (Plate 1).

The Land Rover rounded the rocks alongside our dining tent. At least the rental company had sent a large, commodious one, with three large roof cutouts for game viewing from inside the vehicle. Given the time we would spend over the next two weeks looking for lions on dusty roads, a comfortable vehicle with unobstructed viewing should improve the volunteers' dispositions and amplify their tolerance. As I headed across the parking area at the road's end, I reassured myself with the thought that it was all bound to be an adventure. We had an engrossing project, a spectacular location, and a superbly comfortable camp. How hard could the rest of it be?

One by one, the volunteers peeled themselves from the Land Rover's seats, finally here after one or two days of air travel, an overnight stay in Nairobi, and a six-hour drive. Introductions took place on the fly. First out was Tabitha, the volunteer coordinator from Earthwatch's U.K. office with a passion for Africa and adventure travel—although she was here to evaluate the project at its inception, she had a disarming grin and engaging wit. Next off were Carol and Mary, who between them had assisted in more than twenty Earthwatch projects, making them both experienced field hands and built-in consultants on program and activities; Pete, a environmental policy officer from England and a specialist on Africa and its regional economic development; Jo, from Miami, with a developing interest in science and biology and a deep love of cats; and Dan and Sue, "techies" lately from Silicon Valley, who wouldn't be flustered by the unfamiliar array of equipment we would be using and could help teach the others. Not a word from anyone about the long hot highway or the crowded vehicle or even the late lunch. Not a bad-looking start after all!

After we'd stowed their gear and grabbed that lunch, it was time for the initial briefing. A lot depended on this talk, because all the volunteers expected to be viewing groups of lions before sunset. After all, their expedition briefing had stated: "Your subjects are a unique set of typically maneless lions that you'll observe from vehicles on game drives. Since lions are mainly active at night, expect to work in evening and early morning shifts. After you locate lions you'll keep track of individuals and scan the horizon for prey or other lions. You'll also photograph lion whisker patterns (for identification), videotape behavior, record data and collect scat, hair samples, and habitat data." I needed to explain that we had run into some snags, that this time-honored approach to studying the ecology and behavior of lions had failed us in Tsavo. And I had to do it without eroding their confidence in the project.

"Despite all our preparations, we have only one lion collared at this point," I began, then decided not to sugar-coat the bitter truth, "and we haven't picked up a trace of him in five days." Then the details: Roland, Samuel, and I, assisted by a team of all-star naturalists, KWS staff, and tour operators, had been combing the ranches for the past two weeks. Our trap, baited first with goat meat and then with a live sheep, was spurned by lions and attracted only jackals, hyenas, and a civet. Although we regularly saw lion footprints (pugmarks) in the dusty roadways, we had managed only fleeting and inconsequential glimpses of the cats, never close enough to dart one. Finally, Roland and Julie Thornton, a British endocrinologist helping us discover the mechanism causing Tsavo's lions to be maneless, ran out of time and had to return home, our principal objective still unfulfilled.

But not for long, because the following morning—before their flight even departed from Nairobi's Jomo Kenyatta Airport—we had our lion sighting, up close and personal (Plate 5). Riding in the cab of a Toyota pickup, our ace naturalist and tracker Paul Kabochi had spotted fresh pugmarks crossing the road. As the truck stopped, Paul alighted in the soft sand as he had fifty times before, to determine when and where the lion had passed. No sooner had the door slammed behind him than a large male lion charged, from a distance of perhaps 40 feet! Only the cool-headed response of the KWS rangers alongside me in the back of the open pickup truck saved Paul—one grunted sharply while the other pitched a water bottle across the lion's path, and the lion broke its charge, allowing Paul to speed around the front fender of the truck. Still, there we were, six of us sitting in the open bed, about 15 feet from the lion, still bristling and growling. I don't think I'd managed much more than a dropped jaw....

Thankfully, he wasn't looking for a meal. The lion had been tending a female, who was still visible at the edge of the woods. His charge evidently was motivated more by jealousy and annoyance than by hunger, as mating lions won't eat for days on end. Still, his continued protests, which included running feints and blustering growls, led us to switch vehicles with a film crew that was covering our activities from a safe distance. When we next approached the mating pair, we had tranquilizer darts at hand, the vets in position, collars handy, and six thousand pounds of British steel surrounding us.

We wanted to dart both lions, to secure hair samples and to radio-collar them. Old-time hunters had always maintained that when shooting a group of lions, the male should always be shot first: Females will stay by the body of their fallen leader, whereas males supposedly turn tail and run at the first shot. Francis Gakuya, a KWS veterinarian, successfully darted the male, but as we approached the fallen cat, driving through the thick woodland, the female kept slipping off until she finally disappeared altogether. Certainly, the ranch lions were living up to their reputations as skittish.

The male, however, was waiting for us, crumpled on the ground where the ketamine tranquilizer had finally overtaken him. As the drug did not deprive him of consciousness, only of voluntary activity, we took care to cover his head with a poncho to dull his sensitivity to the activities now going on around him. After the vets checked vital signs, Alex and I proceeded to give the big male a physical exam. We estimated his weight at 350–375 pounds. As I parted his lips to examine the wear on his teeth, an indication of age, his huge head swung up, open eyes staring blankly through the foam that the vets had applied as protection against desiccation. I hurriedly vaulted over his back onto his blind side and helped pull the poncho back over his eyes. Francis readjusted his dosage so that we could complete our examination before the lion was fully recovered. Although the lion's mane was sparse by most standards, this male had a very substantial mane for Tsavo. We took care to gather some hair follicles from each portion of the mane so that Julie could assay their ultrastructure for testosterone receptors. Her studies on humans and red deer had identified a possible mechanism for male hormones to produce manelessness.

The collar we strapped on was a technical marvel. Besides providing a secure nylon collar holding a radio transmitter and battery pack, this collar also had a global positioning system (GPS) device that was programmed to turn on every seven hours, record the lion's coordinates and the ambient temperature of his surroundings onto a memory chip, and then shut down. As Francis injected the male with a ketamine inhibitor, giving us less than a minute to climb back into the vehicle and withdraw to monitor his recovery, I marveled that the entire flurry of data recording had only taken forty minutes. Once the male was fully recovered, we left him alone to find his mate again. At last we had part of what we needed for a successful Earthwatch project.

But that was Sunday morning, and it was now Friday afternoon, and no further lion sightings had been made. I'd headed off to Nairobi and Voi on errands, coordinating logistics and briefing KWS staff—including Tsavo East's newly appointed assistant director, Peter Leitoro—on the project's goals and methods. Meanwhile, Alex and Paul repeatedly attempted to relocate the collared lion but failed to register even the faintest radio signal. Despite prior testing, we wondered whether we had a defective collar or whether our male had simply been a nomad passing through the region.

So the mission I had to give my Earthwatch team on their first afternoon was a very different one from what they'd been led to expect: We had to look *for* lions, not look *at* them. This wouldn't be easy, because none of the lions had yet become accustomed to human observers. And now that Dr. Gakuya and his assistant had been recalled to headquarters, we couldn't simply dart any lions we observed but had to bait and trap them. Small wonder that most scientists choose to study lions in open grassland habitats, where vegetation and *kopjes* don't interfere with radio signals and visual contact can be established once you are within a mile or two of your quarry! But it was the mysteries of Tsavo's lions that had led us all here, and (thankfully) both volunteers and scientists were up to the challenge.

* * * * *

To a person, we had all been drawn to this region and project by the notoriety of Tsavo's lions. A century ago, while the British were laying a railroad from the Indian Ocean coast to Lake Victoria, the railway crews ran into calamity at the Tsavo River. Two male lions there began systematically hunting, killing, and eating railway workers. Over a nine-month period, they claimed an estimated 135 victims before a railway engineer named John Henry Patterson (no relation) finally shot them. Hailed as a liberator by his railway crews, Col. Patterson became a celebrated author when his book *The Man-eaters of Tsavo and Other East African Adventures* was published in 1907. What led the lions to attack and kill so many people? How did they maintain their "Reign of Terror" for so long? Do man-eaters still prowl Tsavo today? The 1996 movie *The Ghost and the Darkness*, inspired by Patterson's book, had reawakened public interest in these questions—millions of people were now familiar with the Tsavo lion story.

Or part of it. Although my colleagues and I couldn't ignore the sensational man-eating aspects of the story, the Tsavo lions had riveted our attention for another reason. Both of the Tsavo man-eaters were fully adult males, but neither of the animals, now on public display at Chicago's Field Museum, had a mane. And a quick reconnaissance expedition to Tsavo in 1998, with Field Museum fellows Chap Kusimba, Tom Gnoske, and Julian Kerbis Peterhans, turned up only three male lions, but all three had been maneless and two were fully adult, suggesting that maneless lions remained in Tsavo. The following year, Roland Kays (then a postdoctoral fellow at the Field Museum) and I spent four months in Tsavo, systematically documenting the park's lions and recording the condition of the manes of males and the social groupings in which they occurred. That fieldwork established beyond doubt that male lions in Tsavo were typically maneless. It also showed that maneless males successfully defend large prides of females and young from rival males. In short, maneless lions weren't losers but rather pride masters. Clearly, another set of rules for lions was in force in Tsavo.

The Earthwatch project was an outgrowth of that work. Although Roland's and my survey had documented five different prides in Tsavo East, each headed by a lone maneless male, we had no idea of how long these males retained their tenure, what size their territories were, or how often their patrols and hostile encounters with other males took place. We didn't know how often males joined females in hunts or what advantage such mixed-sex partnerships might represent in tackling prey like buffalo. But to approach these questions, we needed regular observations of the lions, and only radio collars would offer such regular contact with lions in the dense thorn-scrub woodlands of Tsavo. Collars, we knew, were impossible to use in the national park, mainly for emotional and cosmetic reasons. In Kenya, and all of East Africa, Tsavo is practically synonymous with unspoiled, untamed wilderness, and wildlife lovers everywhere know that the rest of Kenya (and Africa, and the world) grows tamer every day. No one visiting this pristine place wanted to see any of it, especially lions, tamed or controlled. If collars were essential to study the ecology and behavior of these lions, we needed to work outside the park, on the adjoining ranchlands.

Surrounding Tsavo National Parks is a sea of dry savanna and desert scrub. Much of it is too dry for crops and is exploited mainly by pastoralists, herders of goats, sheep, and cattle. This region represents the front line of conservation, where the rubber meets the road, so to speak. Kenya's largest populations of both elephants and lions live in Tsavo, and none respect the 1946 boundaries of the park. Resident animals move freely beyond the park boundaries to graze, browse, and hunt, and their offspring seek homes in the adjacent ranchlands. Not surprisingly, elephants and lions cause enormous damage to people living on these ranches. Knowledge of their behavior and ecology should furnish valuable information for mitigating animal-human conflicts, and this is a central responsibility of the KWS and Dr Kasiki.

Therefore, studying ranch lions could fulfill two goals: to gather information on behavior and ecology that would explain the distinctive appearance and behavior of Tsavo lions, and to collect this information in the environmental context that most often threatens lion survival—depredations on livestock and attacks on people outside of protected areas. We soon realized that a third important goal might also be attainable, once we began to explore the logistics of this project with Steve Turner, a Nairobi outfitter and tour director turned conservationist. Turner was attempting to create a network of independent ranches managed to form a nature conservancy, a protected area that would unite the spread arms of Tsavo West and Tsavo East. Funds derived from our project would help to pay the conservancy's lease on Taita Ranch.

* * * * *

My own interest and involvement explained, it remains to consider why my employer, Chicago's Field Museum,* would pay me to set off on expeditions that don't return with trunks of specimens, artifacts, and other treasures. In part, the search for "booty" represents a classical, antiquated view of museums, now found mainly in "Indiana Jones" movies. Today, most of the world's major natural history museums are bustling centers of modern science, equipped not only with irreplaceable and priceless collections but also with sophisticated imaging equipment, high-speed computing arrays, and bustling biochemical laboratories. The genetic and hormone samples we are collecting from Tsavo lions will be grist for the biochemical mills of the New York State Museum and Bradford University in the U.K., while the location data we collect with the satellite collars will be analyzed with geographic information system software on the Field Museum's computers.

Natural history museums serve as archives of the diversity that makes up our natural and cultural worlds. They are repositories of specimens and artifacts that hail from near and far, from ancient times to yesterday. Whether regional or international in scope, the collections of good museums are encyclopedic. That is, they are comprehensive and represent all branches or varieties of a subject. Even the most active of these collections includes specimens that are infrequently accessed. The wealth of any reference collection lies in its potential applications, not only those that have been realized.

The work of museums is cumulative. Collections must continually grow to stay vital and relevant. Growth inevitably enhances the encyclopedic character of collections—in comparative analyses, bigger *is always* better. In addition, new technologies extract unanticipated insights from existing spec-

*www.fieldmuseum.org

imens but often necessitate adding new collections, such as recorded sounds for sonographic analyses or frozen tissues for molecular techniques. And of course, continual growth requires ever-increasing expenditures of personnel, space, and material resources to service the accumulating collections. The greatest challenge that museums face is to convince the public of the intrinsic value of collections and the need for substantial commitments to maintain and develop them.

Collections lie at the core of any museum—the people who work in museums serve as their stewards and interpreters. Most serve to *make the specimens in the museum's core accessible and intelligible to elements of society that would utilize them.* Exhibitors do this in three-dimensional forms placed on display for the museum-going public. Educators devise formal and informal instructional programs for education at all levels. Curators and other scientists conduct specimen-based scholarly research. Each branch of a museum's staff brings instruction, appreciation, and understanding from the objects under their care to their respective constituencies.

This book represents my attempt to make museum specimens "talk." Every one of the twenty-one million specimens housed at the Field Museum (and at hundreds of others like it) has stories to tell. Many concern the evolutionary histories of plant and animal groups, their growth and development, their form and function, their ecology and behavior. In concert with other specimens and expedition notes, they chronicle communities in distant places and times. Some, like the Field Museum's diorama of Mexican grizzly bears and imperial woodpeckers in a ponderosa pine forest of the Sierra Madre Occidental of northern Mexico, represent species and communities no longer in existence anywhere on the planet; specimens archived in natural history museums are our only means of learning more about these species. However, none of the more than 170,000 mammal specimens in Chicago has a more compelling and involved history than the two lions that Col. Patterson shot in British East Africa more than a century ago. Their story offers an opportunity to synthesize and unify the various disciplines comprised by "natural history" and to communicate this articulated vision to the public. And public education is a principal goal of natural history museums.

Chapter 1
The Reign of Terror: The Lions of Tsavo Attack

The Uganda Railway

Now from the town of Mombasa, a railway line extends into Uganda;
In the forests bordering on this line are found those lions called "man-eaters," and moreover these forests are full of thorns and prickly shrubs. (Roshan, 29 Jan. 1899, in Patterson 1907, *The Man-eaters of Tsavo and Other East African Adventures.* p. 340)

A bit over a century ago, two male lions began eating railway workers in Tsavo, Kenya. But what were the railway crews doing there in the first place? Why should the British build an expensive railway from one remote outpost to another? What imperial objectives could be served? And how costly was the disruption and delay caused by these lions? It is impossible to reckon the place of Tsavo's lions in East African history without considering the railways' history.

Fortunately (for natural historians at least!), the history of the East African railways has been masterfully written. An official, authoritative account, *Permanent Way* (1949), appeared in two volumes written by Mervyn Hill, one devoted to the Kenya-Uganda railway constructed by the British and the other devoted to the Tanganyika railway that was at least initiated by the Germans. Subsequently, Hill's encyclopedic treatment of the railway itself was set in a richer historical and social context by Charles Miller. Miller's very interesting account, *The Lunatic Express* (1971), offers an engaging view of the railway and its unbreakable ties to East African history. Consequently, only the broadest brush strokes need be given to these histories here.

Africa in the Nineteenth Century

The preeminence of imperial Britain during the nineteenth century was undeniable. Its overseas possessions and military and industrial might had grown in tandem. Canada, Australia, India, and other colonies provided Great Britain with ample opportunities for commercial development, natural resources for its industries, and markets for its produce. As Hill put it, for Britain, East Africa was little more than "a guard-room along the sea-way to her Asiatic possessions." However, by the end of the nineteenth century, Africa represented an important opportunity for both territorial and economic expansion for other European powers. Germany had just emerged from the Franco-Prussian War of 1870–1871 both unified and strengthened and was keen to create an empire of its own. France sought to stanch its European losses with further territorial expansion in Africa. Italy and Belgium also aspired to be colonial powers. And quite aside from their own territorial designs in Africa, all sought to prevent the further growth of British power.

British and German interests intersected in East Africa. The entire coastal region had been solidly British through the 1870s, but Germany gained an African toehold in 1885, when a disaffected minion of Sultan Bargash of Zanzibar, a staunch British ally, placed his territory and subjects under German protection. Overlapping and conflicting responsibilities for the European powers multiplied as territorial claims expanded across the continent.

The Berlin Conference of 1884–1885 was organized to orchestrate the European scramble for African possessions. Representatives of more than a dozen nations, including Great Britain, France, Germany, Belgium, Portugal, and the United States, attended the meeting, and the resulting "Congo Basin treaties" were signed in Berlin on 26 February 1885. They not only regulated territorial claims within the Congo Basin, East Africa, southern Africa, and French Equatorial Africa but abolished slavery, protected navigation and trade, and prohibited preferential tariffs.* In 1887,

* Among the provisions were the following:

VI. All the powers exercising sovereign rights or influence in the aforesaid territories bind themselves to watch over the preservation of the native tribes, and to care for the improvement of the conditions of their moral and material well-being and to help in suppressing slavery, and especially the Slave Trade....

IX. ...the Powers which do or shall exercise sovereign rights or influence in the territories forming the ... basin of the Congo declare that these territories may not serve as a market or means of transit for the trade in slaves, of whatever race they may be. Each of the Powers binds itself to employ all the means at its disposal for putting an end to this trade and for punishing those who engage in it.

XXXV. The Signatory Powers of the present Act recognize the obligation to insure the establishment of authority in the regions occupied by them on the coasts of the African Continent

Bargash granted the British East Africa Association a 50-year concession on the African mainland north of the Umba River. A year later, he granted a very similar concession to the German East African Company for territories to the south. This created two "spheres of influence" that were to dominate East African history for the next three decades and shaped what came to be known as the "hinterland doctrine." This agreement recognized the exclusive rights of a Power to influence and control regions adjoining and interior to its coastal holdings. In truth, both the coast and the hinterland were simply chits for Great Britain and Germany to access and commandeer the resources of the rich and fertile Lake region. The fertile region surrounding the African Great Lakes had been a source of natural and human resources for Arab traders for all of a millennium.

Preparing for a Railway

> *In characteristic fashion, the English did not proceed with very great haste. The wheels of Parliament were spoked with cogitation and rimmed with red tape....* (Beard 1988, 103)

Strangely (for those who don't read endnotes), the development of the East African Railroad was integrally intertwined with slavery. For a half-century, European nations had struggled to suppress maritime slave traffic, but interdiction did nothing to alter the ultimate problems of supply and demand. In the nineteenth century, European nations faced the same vexing problems that Americans face today with respect to the illegal drug trade and came to the same conclusion: Slavery could not be abolished through its distribution network but could be eradicated only by eliminating demand.

Toward this end, an international conference was convened at Brussels in 1889, which gave rise to a General Act in 1890 and an Order of Council in 1892. Third on the list of "most effective means for counteracting the slave trade in the interior of Africa" was "The construction of roads, and in particular, of railways, connecting the advanced stations with the coast, and permitting easy access to the inland waters ... in view of substituting economical and rapid means of transport for the present means of carriage by

---

sufficient to protect existing rights, and, as the case may be, freedom of trade and of transit under the conditions agreed upon.

http://web.jjay.cuny.edu/~jobrien/reference/ob45.html

men."* Other European prescriptions to ensure the peaceful and prosperous development of Africa and its peoples were the organization of administrative, judicial, and military services and the establishment of colonial outposts and communication lines.

Lacking government offices in East Africa, the British government sought to fulfill her treaty obligations through the business concerns of the Imperial British East Africa Company. To counter the increasingly aggressive posturing of German interests in East Africa and to preserve British options, the Company eventually established a sphere of influence that was 750,000 square miles in area. This commitment far exceeded any of the company's commercial goals and dwarfed its meager resources, drawing Parliament into subsidizing some of its major projects. In 1891, the Government informed the Company that it would guarantee £1,250,000 for building and equipping a meter-gauge rail line from Mombasa to Lake Victoria. Five years later, the Uganda Railway Act allocated £3,000,000 for the construction of the rail line, but even this additional allocation proved insufficient, requiring further appropriations. By March 1900, the railhead was still far from Lake Victoria, and Parliament was asked to approve a "final estimate" of £4,950,000. The railroad eventually reached Port Florence, Lake Victoria, on 19 December 1901. When the Uganda Railway Committee was disbanded on 30 September 1903, the total cost of the railway had reached £5,317,000 (Hill 1949).

The Railway Route

It is not, after all, a very serious matter to build four or five hundred miles of railway over land that costs nothing. *The Times* for 28 September 1891: p 60

Arab and Swahili traders had refined routes from the coast to the hinterland over centuries of caravan trade. Mombasa was an ideal, protected port, but behind it lay a pair of formidable obstacles: the waterless Taru Desert and the bellicose Maasai, whose reputation for violence was widely known.† To avoid them, Swahili ivory traders relied more heavily on routes

*Preamble to Article 1 of the General Act of Brussels (Hill 1959).

†In 1878, Henry Morton Stanley stated at a meeting of the Royal Geographical Society, "If there are any ladies or gentlemen ... who are specially desirous of becoming martyrs, I do not know in all my list of travels where you could become martyrs so quickly as in Masai" (Miller 1971).

far to the south, now in German possession. Numbering only fifty thousand souls, the Maasai controlled a vast area in southern Kenya and northern Tanzania.

Eventually, European interests in the interior grew to the point that they began to probe some of these neglected routes. In 1883, the Royal Geographical Society funded an expedition by Joseph Thomson, who set off from Mombasa with the goal of crossing Masailand to Lake Victoria. Although the German explorer and missionary Johann Ludwig Krapf had repeatedly crossed it earlier, Thomson was the first to describe the Taru Desert on Tsavo's southeastern border: "Weird and ghastly is the aspect of the grayish-coloured trees and bushes; for they are almost destitute of tender, waving branch or quivering leaf.... The wind... raised only a mournful whistling or dreary croaking, 'eerie' and full of sadness, as if it said, 'Here all is death and desolation!'" (1885, 72–73). He then headed to the Taita Hills, Taveta, Lake Victoria, and eventually to Mount Elgon, all in Maasai-held territory. Through luck and consummate communication skills, Thomson survived his travels, and his trip opened the interior to British interests. His book *Through Masai Land* (1885) is a unique combination of unbridled curiosity, passionate impressionism, and boundless courage. One measure of the man is given by his effortless discovery and effective exploitation of tribal customs: "Spitting, it may be remarked, has a very different signification with the Masai from that which prevails with us or with most other tribes. With them it expresses the greatest good-will and the best of wishes. It takes the place of the compliments of the season, and you had better spit upon a damsel than kiss her.... As I was a *lybon* [medicine man] of the first water, the Masai flocked to me as pious catholics would do to springs of healing virtue...." (1885, 290).

Captain Frederick D. Lugard joined the British East Africa Company in 1890, after celebrated campaigns in Afghanistan, Sudan, and Burma. His first job was to explore the Sabaki River and surrounding countryside and to open a trade route to Machakos that circumvented the Taru Desert. His first expedition lasted five months, but he immediately set out again, this time for Uganda. On that trip, he encountered a Swahili slave caravan near the Tsavo River in August 1890, disarmed the slavers, and sent them under guard to Mombasa. Liberating the slaves, he found some nearly moribund from their travel, and nursed a few of these back to health himself (Hill 1949, Miller 1971).

The following year, Parliament guaranteed funds for constructing a railroad, and in December 1891, a survey expedition left Mombasa to determine and plot its future course. The expedition was perhaps typical of the era in composition and size; it consisted of 7 Europeans, 41 Indians, 7 Swahili headmen, 40 guards, 270 porters, 24 cooks, servants and gun-bearers, and 60 donkeys. One division followed the course of the Sabaki River, which meets the Indian Ocean about 70 miles NNE of Mombasa. The other took the more direct route across the Taru Desert, heading WNW out of Mombasa, and the two met in Tsavo three weeks later. The report of the group crossing the Taru stated, "Across this desert, marching is usually done at night, and water is carried in tins; but even so, men suffer greatly from thirst, and a porter will often barter a whole month's pay for a drink from the supply of a more provident comrade" (Hill 1949, 71).

But the waterless Taru Desert was only one hurdle; other major obstacles lined the caravan route that the railroad followed over the most inhospitable terrain. Beyond lay 200 miles of sparsely settled scrub country, home to tsetse flies, lions, and malaria-bearing mosquitoes. Once the railroad climbed out of the lowlands into more hospitable highlands, the survey crossed the Great Rift Valley and another mountainous escarpment, and then entered marshy lowlands surrounding Lake Victoria. The survey expedition returned to Mombasa in September 1892 and delivered final estimates and plans for the railway to Treasury in March 1893.

The railway's chief engineer was George Whitehouse, who arrived in Mombasa on December 11, 1895. Whitehouse had prior experience constructing railways in South America, Mexico, India, and South Africa, and knew how to adapt crude survey routes to local variation in terrain. But these survey routes posed special challenges that would take all the ingenuity and resourcefulness of thousands to complete. For example, no one relished the idea of laying a route across the Taru Desert, but the alternatives were too costly and impractical. In 1896, the "road" cut by Lugard in 1890 from Makongeni to Tsavo along the Sabaki and Galana Rivers had already become completely obliterated by regrowth. Where the Sabaki meets the Indian Ocean there are substantial fringing reefs preventing a railroad ending there from handling large marine traffic.* Additionally, the Sabaki route would add more than 60 miles to the length of a line based in Mombasa.

And the railroad would have to supply itself, tying up precious human and resources. For most of its length, there was simply no alternative to rail transport. Despite efforts to use draft wagons and pack trains to carry goods, the costs were crippling, because camels and horses quickly succumbed to East African climates. Tsetse flies and lack of water precluded the use of

*Today, Kenya's Malindi Marine National Park protects only one of these reefs, which support extremely diverse Indo-Pacific faunas.

bullocks between Mazeras and Kibwezi. The limited reliability that could be placed on animal transport is shown by Railway veterinary statistics for 1897 and 1898: All 63 of the camels in the railroad's corrals died, as did 128 of 350 mules, 579 of 639 bullocks, and 774 of 800 donkeys (Hill 1949).

Water was a central issue, and along the route selected it was available at only three places between the coast and Mtito Andei (at Mile 161): Maji ya Chumvi ("salt water") at Mile 32, Voi at Mile 100, and Tsavo at Mile 131. Accordingly, perfect coordination was needed between railhead and trains carrying goods, water, and advance parties. Two supply trains were needed daily: one at daybreak with fuel and materials, the other at dusk with food and water. At least ten thousand gallons of water were needed to supply men, animals, and locomotives. Porters and mules supplied the forward parties from the railhead, and the railhead camp shifted with every eight miles of completed track. As Hill pointed out, the availability of supplies limited the size of the advance parties, and this in turn limited the rate of progress that could be expected from the crews.

Water, painfully transported to the camps and stored in tanks, became continually contaminated by use. Inevitably, diseases spread as men in various states of health used unsanitized objects to obtain water from the storage tanks. The workers suffered from ulcers, diarrhea, dysentery, liver complaints, scurvy, and burrowing fleas; nearly half contracted malaria during the first two years of construction. Five or six thousand men depended on the daily water trains, and an occasional derailment became a matter of life and death (Hardy 1965). By March 1897, plate-laying was completed through Mile 38 and survey crews had reached as far forward as Tsavo. Railhead had progressed as far as the Voi River by 28 September 1897 and reached Mile 121—only ten miles from Tsavo—by December 1897 (Hill 1949).

Even this progress would have been impossible without massive immigration of outside laborers and artisans, and colonial India was the obvious source. In agreeing to supply the project, India's government stipulated that at the end of his contract, each worker must be free to remain in East Africa if he so chose. In answer to the need for manpower, nearly 4,000 Indians arrived in 1896, but an outbreak of bubonic plague on the subcontinent interrupted the flow of workers, closing Karachi from April until September 1897. However, by the end of 1899, the number of Indians joining the effort had grown to 18,030. By 1903, a total of 31,983 Indians and Pakistanis had been employees of the railway.

Excepting Swahilis, who were essential in guiding the survey crews and in provisioning supplies, native Africans contributed relatively little to the railway's construction. Their lack of involvement was more economical than political. Because picks and shovels were unfamiliar instruments, natives saw little point in wielding them. Most served as porters, supplying survey crews beyond railhead and carrying goods overland to Uganda. By early 1897, roughly 1,400 were employed, and that number never grew dramatically. In 1899, the Railway Committee lamented that "native labour is of little value, no dependence can be placed upon it, and even famine fails to force the tribes to seek work" (Hill 1949, 192). Even in March 1904, the Railway employed only 3,660 Africans. Truly, the Kenya-Uganda railway was built with British engineering and financing on the backs of Asian laborers.

Railroad Construction in Tsavo

> *We were all very pleased with our new camp, for it seemed an ideal spot.* (Preston n.d., 50, writing of the Tsavo campsite)

The crews at railhead perpetually faced the challenge of the unknown, and one senses that Ronald O. Preston, the railhead engineer, was superbly suited for his role in the construction of the Uganda Railway. Preston had a decade of experience building Indian railroads and took charge of the African plate-laying gangs in 1897. After the Taru Desert ordeal and miles of hacking their way through the thorny acacias and wait-a-bits that choked the area between Voi and Ndi, Preston and his men saw the Tsavo valley as a true oasis. The promise of a stately gallery forest, swiftly flowing streams, and fresh food was there . . . they couldn't foresee the troubles to come.

Few of the men at railhead knew that the name itself was a warning. *Tsavo* means "place of slaughter" in Kikamba (the language spoken by the Kamba people, or Wakamba). It was undoubtedly a reference to the periodic Maasai raids that still swept the region. The vicious onslaught of their attacks was thoroughly corroborated. "Before the advent of British rule, however, sudden raids were constantly being made by them on the weaker tribes in the country; and when a *kraal* [corral] was captured all the male defenders were instantly killed with the spear, while the women were put to death during the night with clubs. The Masai, indeed, never made slaves or took prisoners, and it was their proud boast that where a party of *elmorani* [warriors] had passed, nothing of any kind was left alive" (Patterson 1907, 234–235). Despite its inviting lushness, the old caravan camp by the Tsavo River was utterly deserted when the railway crews arrived.

Even worse, there were also tales of mysterious disappearances. Preston observed that "all the old caravan leaders had disliked this camp for some reason or other, and it was a noted place for desertions, very few caravans passing through Tsavo without a couple of their porters being missed. The idea was that they cleared off back to the coast, by the Sabaki slave-trade route, which was a very fair path running down along side of the Sabaki River all the way to the coast. A strange feature of these desertions was that even on the return journey from up-country, the porters would seemingly desert, leaving their loads in camp; but stranger still, some of the men who had no loads to carry would make themselves scarce" (n.d., 53). What man who has toiled for weeks would desert his job just before payday?

Railhead had hardly reached the west side of the Tsavo River by means of a temporary wooden bridge when Preston learned that one of his men had disappeared. Organizing a search on the riverbank near his abandoned loincloth, they soon uncovered the unfortunate man's remains. "The skull and the feet were untouched, but all the flesh had been torn from the body. We had not far too [*sic*] look for the cause of the tragedy, paw marks of lion being easily seen all round the remains.... So here was the Tsavo mystery solved at last, this man-eating brute was responsible for all the missing porters from the caravans, and was also the object around which centered all those weird tales of the camp" (n.d., 54). Preston was not a squeamish man, but the dispiriting effects of the remains were undeniable: "I have witnessed many an accident with fatal consequences, in some of which the unfortunate subjects have been badly mutilated, but the sight of this skeleton, from which the flesh had been revenously [*sic*] torn was one of the most gruesome spectacles imaginable. We buried the remains at the spot, and on returning to camp set all the men to putting up thorn bomas [brush stockades] round their tents" (n.d., 54).

The following day, Preston and his men searched in vain for the lion. Although they saw no trace of the lion, they stumbled upon a number of additional human remains, skulls and parts of skeletons, confirming that yesterday's killing was neither the first nor likely the last. When a second man was lost to the lion, panic mounted: "These two visits fairly demoralized our workmen, who were disinclined to work; but when I explained to them that this man-eater probably confined its attention to this particular camp by the river, it was up to us to get a move on and lay rails as quickly as possible so that we could move camp beyond its haunt. The men evidently saw the wisdom of this and worked like Trojans till we got the rails sufficiently advanced to lay another siding a few miles away from this death hole, and within a few hours of laying the siding Tsavo was deserted by us. We were generally fairly quick about shifting camp, but this move from Tsavo was certainly a record one, so far as time was concerned" (n.d., 58). It was not until the railhead reached Kenani, about 20 km from Tsavo, that they escaped the range of this man-eater.* Preston's tally of the men taken by lions prior to the move of railhead in April 1898 was "16 Punjabi workmen and one Punjabi head-man" (p. 64).

Preston returned to Tsavo some time after the railhead had moved west. He took the trolley back to Tsavo and arrived very early one morning. As he pulled up to the siding, he noticed a man waving frantically to him from atop a tree, urging him to approach. The man's tent lay in ruins a few feet from the tree. On his arrival, the man, a Greek contractor, explained that a lion had sprung on his tent in the middle of the night, seizing the mattress he had been sleeping on. As the lion carried the mattress off, the man slipped off and scrambled up a tree before the lion realized his mistake. The man had stayed treed all night, awaiting daylight. The tent itself was torn to shreds by the lion, either in the initial attack or afterwards† (n.d.).

"The Man-eaters of Tsavo"

Tsavo is today a name known around the world because of John Henry Patterson and his celebrated book *The Man-eaters of Tsavo and Other East African Adventures,* first published in 1907 and still in print today. Patterson was then a civil engineer and lieutenant colonel who had just finished service in India. In 1898, he was assigned the task of completing the permanent rail fixtures within 30 miles of Tsavo Camp, including the permanent bridge work over the Tsavo River. To speed construction, plate-layers at railhead constructed temporary bridges from timbers—Patterson's task was to build a bridge over a durable foundation of stone. He reached Tsavo by special train from Mombasa on 8 March 1898, a week after arriving in East Africa. He regarded the region north of Ndi as "beautifully wooded country" after the sterility of the Taru, but his enchantment with the area lasted only until dawn. "My first impression on coming out of my hut was that I was hemmed in on all sides by a dense growth of impenetrable jungle. ... I found the whole country as far as I could see was covered with low stunted trees, thick undergrowth, and 'wait-a-bit' thorns" (1907, 16). The only clearings were rhino trails and those made by men for the rails (see Figures 1, 2, 3).

*Earthwatch Team II of 2002 observed the big male lion named "Romeo" at locations twelve miles apart on successive nights on the ranchlands SE of Voi.

†It is curious that Patterson included in his account this same story, but without reference to Preston. Patterson gave the man's name as Themistocles Pappadimitrini and amended his story to note that, having survived this attack by the lion, he would shortly die of thirst on a trek across the region's wastelands.

Figure 1. Lt.-Col. J. H. Patterson in front of his tent at Tsavo Camp. (Field Museum neg. # Z-94092, © The Field Museum.)

Figure 2. Machan or shooting platform used by J. H. Patterson to shoot Tsavo's "second man-eater." (Field Museum neg. # Z-93657, © The Field Museum.)

Figure 3. Dense thorn-bush lining the tracks near Tsavo during the "Reign of Terror." (Field Museum neg. # CSZ-48827, © The Field Museum.)

He had just drawn up his plans and requisitioned supplies from Mombasa when the first men turned up missing. His statement that "one or two coolies mysteriously disappeared" might refer to the same attacks that Preston described. Railhead had just reached the west side of the Tsavo River, and Tsavo Camp was still home to nearly three thousand men scattered around it. Patterson simply relayed what he was told about the victims, including "I was told that they were carried off by night from their tents." He doubted the veracity of the man-eater story, believing that the men were instead victims of foul play over unspent wages.

But the presence of man-eaters was brutally confirmed shortly afterward, with the death of Ungan Singh, a *jemadar*, or Indian leader. A tentmate of the unfortunate *jemadar* confirmed that, around midnight, a lion had stuck his head in the open door of the tent and seized the throat of the man lying closest to the entrance. The man uttered a single cry and was carried off, his heels creating furrows that marked the direction he was taken. The following day, Patterson and the veterinarian Capt. J. A. Haslam discovered the man's remains, his body torn to pieces and largely consumed while his head remained intact. Pierced by giant tusks, his face had a horrified look frozen on it. Pieces of his body had the skin licked off and the tissues apparently sucked dry of blood. During the next two or three days, the camps lost two more victims from the camps. Each time, Patterson was out waiting for them with his rifle, too far to see or shoot but close enough to hear the frenzied cries and shrieks of his workmen.

Patterson's book includes various accounts of lion attacks on workmen that must be read to be fully appreciated. Given that his authoritative and engaging narrative is still available, it would be folly to identify all of them, or even the principal ones, here. Instead, I endeavor only to relate enough historical facts to orient subsequent discussions of man-eating and to clarify his points of difference with Preston.

Until the attack on Singh, the work camps had been strung for miles on either side of Tsavo. Few, including Patterson's, had been fortified with thorn *bomas* to protect them, despite Preston's prior experience. But consolidation of the camps and construction of massive barriers of two-inch thorns did little to slow their attacks. Patterson would spend the night perched on a *machan*—an elevated shooting platform much used on tiger hunts in India—at the scene of an earlier attack. Sometimes he used livestock as bait, other times the corpses of transport animals that had succumbed to sleeping sickness (trypanosomiasis) transmitted by tsetse flies. Too often, while sitting in wait, he only heard the shrieks and cries of another attack reach him in the blackness of the night from a nearby area.

To multiply his chances of killing the lion or lions, he began lacing the corpses of livestock with strychnine, but he found that the camp's scourge "much preferred live men to dead donkeys" (Patterson 1907, 106). He also sat up all night in a "goods-wagon" (boxcar) stationed near the site of the previous evening's attack but managed only to elicit an ambush and charge that was turned at the last minute. He was simply eluded at every turn, finding only tattered clothing, gruesome remains, and harried men as the lions' calling cards.

Patterson's narrative for the middle third of the year 1898 is filled with descriptions of the work crews, life in camp, hunting sojourns, and the like. These adventures conceal an important fact: From late April until November, there were no more lion attacks on the work crews at Tsavo. Patterson attributed this to having given the lions "a bad fright" from the goods-wagon, but repelling dauntless predators for six months by such an encounter seems far-fetched. During this respite, he believed the camp's nemesis moved up rail, to interfere with plate-laying operations beyond Tsavo, claiming victims at railhead and about ten miles away near Ngomeni. Concomitantly, Patterson constructed a lion trap from railroad sleepers, tram rails, chains, and wire, one secure enough to admit humans as bait.

A bounty on lions along the railroad had been instituted in July 1898, apparently in direct response to the attacks that took place at Tsavo and beyond. According to Hardy, a 200-rupee reward was offered for "the skin of any lion shown to the satisfaction of the Managers to have been destroyed within one mile on either side of the Railway line and to a distance of five miles East and West of the River Tsavo" (1965, 151), but in *The End of the Game,* Peter Beard stated, "For years, the prize of 100 rupees was awarded to anyone who shot a lion within one mile of the "Permanent Way"" (1988, 103). The bounty attracted many would-be hunters to Tsavo, but few stayed for long. The constant persecution by trigger-happy sportsmen patrolling the tracks may have helped to keep lions at bay, accounting for the hiatus in Patterson's log.

The attacks resumed again, apparently in November, when vigilance in the camps had subsided. Camps had again become scattered along the line near various permanent workings, and workers again slept unprotected in the open to escape the drought's oppressive heat. Now the lions were bolder and scarcely retreated from the *boma*s before beginning their meals in full view of the horrified men. Shockingly, they continued eating men despite frantic gunshots or torches being hurled in their direction. Because of the numbing repetition of the attacks, nightfall brought a sense of foreboding as the hunt-initiation roars of the lions seized everyone in camp with dread, including Patterson. When the roars quieted and the lions approached, the men shouted from camp to camp "*Khabar dar, bhaieon, shaitan ata*" (Beware, brothers, the devil is coming).

Patterson's persistence in hunting the man-eaters finally paid off in December 1898, when he succeeded in killing both lions that had been seen in camp. The first was shot on December 9, from a 12-foot *machan* precariously balanced in the thornbush; one bullet struck him in the heart, the other in the thigh. The lion proved to be quite large, recorded by Patterson as 9 feet 8 inches long and 3 feet 9 inches high. "The only blemish" noted on the "first man-eater" was that the skin was much scored by the thorns of *bomas* he had lately penetrated in his depredations on the camp.

The second lion next attacked a few nights later but failed to kill the Permanent Way inspector he had targeted. Patterson then fashioned a blind in an iron shanty, tying three live goats to a 250-pound rail; just before daybreak, the lion seized one of the goats and dragged all three plus the rail off into the bush. Catching up with him after daybreak, Patterson succeeded in firing both barrels into his shoulder at close range, but the lion escaped, wounded, and wasn't seen again for ten days. The second man-eater treed a number of workers on December 27, again failing to kill anyone but forcing his way into a number of tents. The following night, Patterson placed a *machan* in the same tree and, after firing a fusillade of shots (at least six of which struck home), managed to bring the lion down. The second lion reportedly taped at 9 feet 6 inches long and 3 feet 11½ inches tall.

Patterson's Character

In *The Iron Snake,* Ronald Hardy interpreted lack of acknowledgment between Patterson and Preston as the result of active jealousy between the two men, guessing it was founded on class distinctions and personal mannerisms. Whereas Patterson was supposed to be high-mannered and aristocratic, Preston was raised in an orphanage; whereas Patterson hunted for trophies, Preston shot game for the tables; and so on. In virtually all of these comparisons, Preston emerges as a sort of blue-collar hero, and one cannot read Hardy's account without somehow cheering for Preston's matter-of-fact heroism. Hardy used descriptions such as "arrogant," "a martinet," and a "product of rigid hierarchies of caste and rank" for Patterson. Even his loyalty to the railroad crew is branded "patriarchical" (Hardy 1965).

A careful reading of *The Man-eaters of Tsavo* does substantiate that Col. Patterson (Figure 4) routinely used a number of authoritarian measures whose descriptions are woven into Hardy's indictment. He recorded the transgressions and errors of his men in a pocket notebook as these were spotted during his patrols. He held public "trials" at noon to bring "evildoers" to justice, and exacted fines for all manner of misdemeanors.* And he abrogated a contractual agreement that a number of men had reached with the railway before they emigrated to East Africa. The pay for stonemasons was set at 45 rupees per month and that for common, unskilled workers at 12; naturally, many recruits hired on as stonemasons. To counter this, Patterson instituted a system of piecework in place of the flat monthly salary they had been promised, which was widely perceived as punitive. However these management practices may have expedited the construction of the Tsavo Bridge, they unquestionably precipitated the mutiny that took place on 6 September 1898. The mutiny was suppressed only with the aid of regional police reinforcements.

In truth, Tsavo Camp must have been a personnel manager's nightmare. Patterson's crew was composed of distinct, sometimes intolerant, ethnicities, each with its own customs, standards, and expectations. Religious

*His summary treatment of Karim Bux, whom Patterson suspected of feigning sickness, is illustrative. Patterson actually set his bed afire to prove that he could get out of it!

Figure 4. Portrait of Lt.-Col. J. H. Patterson. (Field Museum neg. # CSZ-49222, © The Field Museum).

tensions between Muslim and Hindu workers simmered, and local workers frequently deserted. Two plots to murder Patterson were averted by the loyalty of some of his workers, who revealed the plans to him beforehand. After shooting the man-eaters, he declined the gift of hundreds of rupees that the workers had collected to thank him for his heroism. Eventually, he accepted a silver bowl inscribed with their grateful thanks, which became his most treasured trophy, more even than the man-eaters themselves. The epic poem he was given, written in Hindustani by the *mistari* (or superintendent) Roshan, is certainly a tribute fit for a bona fide hero.

Hardy's unflattering characterizations of Patterson in Tsavo are difficult to reconcile with what is known of his later life. During the Gallipoli campaign of the First World War, the British sought to take the Dardanelles from the Turks. Patterson, now a full colonel, led a brigade of the Jewish Legion. He characterized his legion as "makeshift" and composed mainly of London tailors, "who had never wielded more deadly weapons than needles and shears in their lives" (Carlozo 1996). But out of this group came such Zionist luminaries as David Ben-Gurion, who later became Israel's Prime Minister, and Jacob Epstein, the acclaimed American Expressionist sculptor. In his introduction to a modern edition of Patterson's classic book, big-game hunter Peter Capstick related a signal incident. Patterson's unit was being inspected by a brigadier notorious for his anti-Semitism. Inevitably, the general found fault with the bearing and demeanor of one of the men, pronouncing him a "dirty little Jew." Without a thought to his own military advancement (in fact, at almost certain cost to it), Patterson instantly ordered his entire brigade to fix bayonets and about-face, surrounding the bigot with a thousand invitations to death by impalement, and demanded the general offer the private a personal apology. Either Hardy's characterizations are wildly off base or Patterson learned a lot about men in the ensuing decades. In any case, Patterson's numerous accomplishments surely warrant a more detailed biography, and the complex reactions he elicited in his associates promise great interest.

The Human Toll

An accurate tally of the lions' depredations is impossible, because detailed records were not kept. The two principal authorities, Patterson and Preston, took pains not to cite one another. This is probably the factual basis for Hardy's dramatic contrast and inferred rivalry between the two men. Some of the attacks they describe are sufficiently distinctive in details that they can be unambiguously identified in both accounts. Others are sufficiently sketchy that they may be uniquely presented in one or the other's book. Consequently, one can neither sum the depredations in both books nor view the fewer attacks reported by Preston as merely a subset of those included in Patterson's account.

Patterson's estimates of the lions' toll varied between his principal accounts. In his 1907 book, he wrote of the man-eaters, "...they had devoured between them no less than twenty-eight Indian coolies, in addition to scores of unfortunate Africans of whom no official record was kept" (107). In the summary he later prepared for the Field Museum, he stated, "... there two ferocious brutes killed and devoured, under the most appalling circumstances, one hundred and thirty-five Indian and African artisans and laborers employed in the construction of the Uganda Railway" (1925, 1).

Hill's official Railway history registered that these two lions "had devoured 28 Indian coolies, and scores of Africans of whom, rather callously, no official record was kept" (1949, 174). Why the uncertainty regarding the deaths of so many Africans? Hardy offered the likeliest explanation: "Natives were employed by the U.R. [Uganda Railway] in simple jobs like water-carrying and wood-cutting and bush-clearance. But they were never on the Roll & they had no names, they were paid in wire or knives or food, a lot of them died at Tsavo but nobody knew how many or cared for that matter...." (1965, 199).

If we accept Patterson's second tally of the human toll claimed by the lions, it is an astonishing amount of carnage. Over nine months (about 270 days), *the lions would have killed 135 people,* averaging a human victim every other day. The lions were rarely permitted to make full use of the victims, with rescue and reconnaissance parties in hot pursuit at first light. With several hours until dawn, two hungry lions might be able to consume as much as 80 to 100 pounds of meat and organs from a human carcass, translating to at most 20 to 25 pounds of flesh each daily. From this volume of food and a lion's estimated energetic demands, it could be inferred that humans were their principal prey throughout this period.

However, no one knows what the lions were eating or where they were during the long lull, from May into September, in their "Reign of Terror." This time period represents the "long dry season" in southeastern Kenya. During the dry season, Tsavo's resident lions typically stay close to permanent water sources, as John Hunter observed and Roland Kays and I recently confirmed (Kays & Patterson 2002)—finding food is simplified at this time because prey species depend on a handful of dependable water sources for drinking. In late 1897 and 1898, the rains had failed completely, making Tsavo exceptionally dry and triggering a regional drought and famine. Mass mortality from starvation was rampant in adjacent Taita and Kamba villages. As discussed in Chapter 3 on "Killing Behavior and Maneating Habits," it seems likely that these committed man-eaters would have continued to eat people—probably exclusively Africans, some almost certainly corpses when the lions first encountered them—from villages along the nearby Athi River. The Tsavo River bridge is just upriver from the Tsavo's confluence with the Athi River.

From Book Legends to a Museum Diorama

After he killed the lions, Patterson's fame carried far beyond the railroad. It seems now a remarkable coincidence that Patterson would choose to write the initial account of his feats of derring-do in the sporting magazine named *The Field* given that this is also the vernacular name of the museum that would eventually acquire his man-eaters. Patterson's article triggered international interest. *The Spectator* of 3 March 1900 stated, "If the whole body of lion anecdote, from the days of the Assyrian Kings until the last year of the nineteenth century, were collated and brought together, it would not equal in tragedy or atrocity, in savageness or in sheer insolent contempt for man, armed or unarmed, white or black, the story of these two beasts...."

Patterson's reputation as a sportsman and naturalist was greatly enhanced by the endorsement he received in his book's foreword from the acclaimed hunter and explorer Frederick Courteney Selous. Selous stated, "A lion story is usually a tale of adventures, often very terrible and pathetic, which occupied but a few hours of one night; but the tale of the Tsavo man-eaters is an epic of terrible tragedies spread out over several months, and only at last brought to an end by the resource and determination of one man" (Patterson, 1907, x). A year later, Teddy Roosevelt hosted both Patterson and Selous at the White House and had them arrange his famous Kenyan safari. In his book *African Game Trails*, published in 1909, Roosevelt described Patterson's work as "the most thrilling book of true lion stories ever written" (12).

Patterson's fame was enduring, and he resumed his lectures after years of service during the Great War. On an American lecture tour that also carried him to New York, Ohio, and Detroit, Patterson gave a lecture entitled "The Man-eaters of Tsavo" at Chicago's Field Museum on November 29, 1924. During his visit, Patterson told the museum's president, Stanley Field, that the two lions he had shot in Tsavo were still in his possession, as rugs with accompanying skulls. Immediately after his lecture, Patterson offered to sell his man-eating lions to the Field Museum. The museum's archives includes his letter, quoted here:

[The Chicago Club] 1st Dec, 1924

To D. C. Davies, Esq.
Director
Field Museum, Chicago

Dear Sir,

Since I have had the privilege of going through the Field Museum and admiring all the rare and beautiful things gathered from the ends of the earth and so artistically displayed in that magnificent building, I have been saying to myself "Since you must part with your man-eaters, get them into 'The Field' if possible."

I can truly say that I would rather they were in your museum than any other I have seen the world over.

I want to dispose of the Man-eaters owing to financial losses which took place during the war, and as I have shown them on the screen and lectured about them to a large and appreciative audience in the Field Museum, it has occurred to me that you might like to have them.*

They are the most famous lions in history, and the late President, Theodore Roosevelt, wrote that there was nothing to equal them since the days of Herodotus.

The man-eaters are huge maneless bush lions and I am certain that they would prove a great center of interest in the Museum.

They are in perfect condition and the price I ask is five thousand dollars.

I am,
Dear Sir, Very truly yours,
J. H. Patterson

The skins and skulls of the two lions were received at the Museum and were examined by an appraiser on January 28, 1925. They were formally accessioned on February 5, 1925, treated as a gift-in-kind credited to the Museum's president, Stanley Field. The Museum wasted no time in getting them mounted and on exhibit. The account of the specimens

*Patterson missed a major economic opportunity when he declined a grant of several thousand acres of land in British East Africa, as a reward for serving his country. At the time, "I was too much on the go to care about it. But I'd have been many times a millionaire today if I'd taken it" (Carlozo 1996).

in its Annual Report states: "They had been preserved for a number of years and were not prepared originally with a view to museum exhibition. Therefore they offered unusual difficulty to the taxidermist and were mounted only by the exercise of much painstaking care and skillful manipulation. This was accomplished by Taxidermist Julius Friesser with the assistance of Mr. H. C. Holling..." (Davies 1926). A photograph of the finished diorama (Plate LXV in the original report, reproduced here as Figure 5) appears in the report. During this period, the Museum also requested that Patterson write an abridged account of the man-eaters story, which it published in 1925 with an initial press run of 6,000 copies. The 40-page leaflet *The Man-eating Lions of Tsavo* originally sold for 50 cents but now costs $4.50.

Although Capstick described the museum's man-eaters as "partially moth-eaten," in reality they are still in excellent condition. One who has only seen pictures of them might dismiss them as threadbare, but even in life, the lions' coats were scarred by thorns and short-furred. Scrub-dwelling lions (and other animals) lack the dense coats and fine condition of their

Figure 5. The man-eaters of Tsavo in 1925, newly mounted as a diorama by taxidermist Julius Friesser with the assistance of H. C. Holling. (Field Museum neg. # CSZ-50965, © The Field Museum.)

relatives from the high plains. In addition, the faces of the man-eaters lack the intensity that characterizes those of living cats, but this is true for virtually all taxidermy on cats. Among the dozens of mounted cats on exhibit at the Field Museum, there are only a few that do full justice to those species in life. But perhaps the most disappointing aspect of the lions to museum-goers is their size: they simply aren't the towering giants described in Patterson's book.

What can the mounted skins tell us of the lion's size in life? The skin of most large mammals is thick and composed of both epidermis (the layer containing hair and glands) and dermis (the underlying layer of connective and supportive tissue). Excluding hair, the skin of a lion is perhaps a half-centimeter (quarter-inch) thick and completely covers its body. To make this three-dimensional form lie flat—that is, to assume the two dimensions that we require of rugs—necessitates that it be trimmed. Friesser, the taxidermist, began his work with flat rugs, which needed to be draped around three-dimensional forms. Again, this dimensional transformation would entail trimming. These processes would make the lions smaller than they were in the flesh. On the other hand, tanned skins often become stretched as they are worked for suppleness. The mounted skins offer only ambiguous and unreliable clues to the lions' size.

Although Patterson reported measurements for both of his man-eating lions in the field, there is reason to question their accuracy. Adult male lions seldom exceed nine feet in length, yet Patterson reported lengths of nine feet eight inches and nine feet six inches and heights of three feet nine inches and three feet eleven and a half inches. To offer some scale, among 150 lions measured by Col. James Stevenson-Hamilton while he was warden of the Sabi Game Reserve (forerunner to South Africa's Kruger National Park), only one even approached ten feet in length (Astley Maberly 1963). Patterson's measurements have fueled impressions that the historic man-eaters and other lions in Tsavo are larger than "normal" lions, and the region's tour operators and visiting journalists have helped broadcast those impressions. No systematic analysis has been conducted (but see Chapter 5).

Few have noted that, in his original book, Patterson recorded a male lion as big as either man-eater that he shot on the Athi Plains southeast of Nairobi; Patterson's Athi lion measured more than nine feet eight inches in length (1907). Because this area is not known for the exceptional size of its lions, it seems likely that Patterson took his measurements along the curves of the neck, back, and tail, whereas official lengths are taken between uprights placed at nose and tail tip.

The skulls of the man-eaters provide another estimate of their size, one unaffected by shrinkage, trimming, exaggeration, or customs of measurements. Analysis of skull measurements is presented in Chapter 5, in the discussion of geographic variation and differentiation of lions.

Other Lion Attacks on the Railway

After the tragedy of Tsavo, many forward gangs insisted on returning by rail to the main camp each night to avoid becoming lion victims (Guggisberg 1961). Only three months after the Reign of Terror ended in Tsavo, lions killed a former customs official named O'Hara, who was charged with building a railroad from Taveta to Voi—both Preston and Patterson described this incident. In March 1899, O'Hara was near Ndi, twelve miles beyond Voi, in a tent with his wife and two children. During the night, a lion entered their tent and seized O'Hara by the head, killing him instantly, and dragged him from bed without immediately rousing Mrs. O'Hara. On awakening, she found her husband dead on the ground outside the tent and the lion standing two feet from her (Hill 1949). She began screaming and was rescued by *askaris* (guards), who also reclaimed O'Hara's corpse. However, the lion besieged the tent all night and was held at bay only with frequent shots from the *askaris*' rifles.

In 1900, Charles Henry Ryall, superintendent of Railway Police, became, in Charles Miller's words, "the first and only official in the history of labor arbitration to be eaten by a lion" (1971, 430) He was only twenty-five years old, and he had just finished resolving a crippling labor dispute in Mombasa. Under his command, the Railway Police had just reined in rebellious workers forcibly, with a baton charge; issued a Riot Act; and jailed the strike's ringleaders in Fort Jesus. Ryall was then summoned up line by additional problems in Nairobi, but he never reached that destination. At Kima Station, 260 miles from Mombasa and only 69 miles from Nairobi, he was attacked and killed by a lion.

This incident differs from many others in having several eyewitnesses, and their accounts were recorded by Maj. Robert Foran in *East African Railway* (1961). Foran interviewed the two survivors of the attack, who had been Ryall's guests in his personal carriage during the lion-turned-man-hunt, as well as investigating officers. Most of the following is drawn from his hard-to-find account.

In 1900, Kima Station consisted of a building fifteen feet square that served as both office and residence for the stationmaster and an adjacent building housing the rest of its staff: a signaler and two pointsmen. The station was surrounded by uninhabited savanna that was well stocked with game, and neighboring stations (Simba and Kiu) were twenty-nine and nine miles distant from it. Lions were much in evidence. One had jumped on the roof of the station and tried to tear away the corrugated iron roof in an attempt to reach the people cowering inside. Shortly afterward, a lion killed and ate an Indian pointsman, a African railway employee, and a number of other Africans.

The lion began to hunt during the day, causing panic among the rail workers and attracting a series of colonial rail officials who all tried and failed to dispatch the animal. An engine driver holed up in the water tank, using it as a hunting blind, but the lion knocked the tank over and tried to drag the man out through the circular hole in the top. The terrified engine driver finally scared the animal away with a shot from his rifle. A crossing loop of rails at Kima Station allowed trains plying the single-track railroad to pass one another there. However, after the loss of the first pointsman, the survivor refused to set the points after dark, forcing trains to set points for themselves. And the lion became bolder, "continuing with its deadly work of destruction of human lives" (Foran 1961, 4).

Ryall was an avid hunter in both India and East Africa and decided to break up his trip from Mombasa to Nairobi with a stop in Kima. Traveling in his private coach were Parenti, a trader and the Italian vice-consul in Mombasa, and Huebner, a German merchant, both headed for Uganda and barred from riding on the construction train. On arriving at Kima, Ryall had his car shunted to a siding about twenty yards from the station building. The three men occupied a compartment with a sliding door opening onto the rear of the car, which consisted of a small platform with steps. A small lavatory compartment bordered this chamber on the other side. Beyond this, at the other end of the car, Ryall's cook and bearer shared a chamber similar to his own.

Ryall's plan was that the men sit up at night, taking turns waiting for the man-eater to appear to investigate the open door to their coach. They drew lots for the three watches: Parenti until midnight, Ryall until 3 am, and Huebner until dawn. Nothing happened during Parenti's watch, and Ryall relieved him. Parenti curled up on the floor, under Huebner's upper berth on one side of the compartment, while Ryall stretched out on the lower berth facing them, after opening the windows to alert him to the slightest sound. He remarked then to Parenti that he could see the eye-shine of some rats over by the landing; inspection of pugmarks the next morning showed that Ryall had actually seen his destiny, intently watching them from a distance.

After the lights were out and sounds subsided, Ryall apparently dozed. The lion stealthily approached the car and stepped onto the rear platform. In a single movement, it rushed into the car, seizing Ryall while actually standing on Parenti's chest. No outcry was heard from Ryall—it is probable that the lion's canines had either pierced his skull, killing him instantly, or that it had broken his neck or crushed his larynx in seizing him by the throat. Parenti, pinned to the floor by the lion's weight, nearly retched at "the nauseating stench of the man-eater" and may actually have passed out. During this commotion, Huebner rolled off the upper bunk and took refuge in the interior lavatory compartment, unheroically bolting the door behind him.*

As the lion turned to leave the carriage with Ryall in his jaws, its weight tilted the spring-loaded carriage, causing the sliding door to close. Trapped inside, the lion must have considered its position for some time, judging from the pool of Ryall's blood found next to the door in the morning. When Parenti regained consciousness, he saw the lion with Ryall's body gripped in his jaws, squeezing his way out through the opened windows. He never thought to use the loaded rifle on the floor next to him, terror rendering him incapable of any action. The lion lighted on the platform and carried Ryall off into the bush.

R. M. Howard was the guard of a goods train that arrived at Kima shortly after Ryall was taken. Arriving at dawn, he found the entire station in helpless panic and confusion; no efforts had been made to deal with the man-eater or even to recover Ryall's body. A small, armed search party was organized and only ten to fifteen minutes later fired two shots to indicate to others waiting at the platform that Ryall's body had been found, the lion apparently having retreated on their approach. Ryall's body had been disemboweled and his intestines dragged some distance away. Despite the passage of several hours, relatively little of his body had been consumed. The remains were gathered, transported that day to Nairobi, and interred in Hill Cemetery.

In the wake of Tsavo, this tragedy stirred the country to action. George Whitehouse, the Railway's chief engineer, offered a reward of fifteen pounds for every lion killed between Nairobi and Makindu, a distance of 199 miles. And Ryall's mother posted a hundred-pound reward for the "Kima Killer." A number of lions were killed in this area, five of them close to Kima, but none proved to be the man-eater. In his article, Foran claimed

*Huebner later dined with Teddy Roosevelt at the Mombasa Club during the latter's famed safari (Roosevelt 1999).

"a man-eater's stench proclaims it for being what it is, so there is little room for error" (p. 8). Perhaps he based this observation on Parenti's firsthand account. It anticipates a discussion of dental injury and disease undertaken in Chapter 4.

The depredations continued. Two railway workers, Costello and Rodrigues, sought the reward for bringing the man-eater to justice and placed a trap baited with a live calf in the bush within a mile of Kima Station. Two months later, the lion's attacks on people ended when he was finally captured, held for some days "to reassure the coolies," then shot. While in the trap, he was photographed, and Costello later showed the pictures to Foran, who stated, "The man-eater appeared to be an unusually large brute, well-maned, powerful and handsome, and patently in prime condition; and, in view of all this, there was no justification for its man-eating proclivities as vast herds of game animals abounded all around Kima for its sustenance. Exactly the same can be said for the notorious man-eaters of Tsavo...." (1961). For some time after Ryall's death, the Railway authorities offered a 200-rupee bounty for every full-grown lion shot within the railway zone (Hill 1949).

The desperate events at Kima were staged and re-enacted during the 1950s filming of the movie *Permanent Way*. As of 2001, Ryall's train carriage remained on public display at Nairobi's Railway Museum, and lion claws said to be those of the "Kima Killer" were on display under glass there. The lion's mounted head was supposedly given to the Nairobi Railway Station by Mrs. E. H. Preston of Nakuru, a relative by marriage of R. O. Preston. In September 1998, Chap Kusimba, Tom Gnoske, Julian Kerbis Peterhans, and I made an assiduous search of the Nairobi Station and interviewed railway officials but failed to find even memories of this lion head. Lions continued to attack residents and travelers after completion of the railroad. Shortly after arriving in the colony, the famous game warden Blayney Percival had come across two victims of a severe lion mauling, Lucas and Goldfinch, and helped to get them hospitalized. Goldfinch had been attacked and Lucas had tried to pull the lion off him, losing most of his hand and stomach in the process. He died, according to Peter Beard, and was one of the first six men buried in Nairobi cemetery, all victims of lion attacks. In *Life-Histories of African Game Animals*, (1914) Roosevelt and Heller described another harrowing event that took place near Machakos. A white traveler was seized and removed from his tent by a man-eater, but the man's companions rescued and recovered him, dressed his wounds, and put him again to bed. No sooner was he left alone than the lion returned, again forced his way into the tent, and carried him off and ate him.

For years, besieged stationmasters along the railway line telegraphed incoming trains regarding the presence of lions in the yard; then points were left unplaced, tanks were left unfilled, passengers were not accepted. In *Simba*, Guggisberg (1961) recounted a number of examples, these from Simba Station to the northwest of Tsavo:

"Simba, 17.08.05 1 hr 45 min. The Traffic Manager: Lion is on the platform. Please instruct guard and driver to proceed carefully and without signal in the yard. Guard to advise passengers not to get out here, and be very careful when coming in office.

"Simba, 17.08.05 7 hr 45 min. The Traffic Manager: One African injured at 6 o'clock again by lion, and hence sent to Makindu Hospital by trolley. Traffic Manager please send cartridges by 4 down train certain.

"Simba, 17.08.05 16 hrs. The Traffic Manager: Pointsman is surrounded by two lions while returning from distant signal and hence pointsman went on top of telegraph post near water tanks. Train to stop there and take him on the train and then proceed. Traffic Manager to please arrange steps" (1961, 204).

Gradually, denser settlement along the railroad and better, more automated maintenance of the roadway lessened the frequency of these encounters and the severity of their outcomes.

Chapter 2
The Terror Continues: Man-eating Lions Today

Dealing with the occasional outbreak of man-eating lions was routine work for all the Game Rangers. (G. G. Rushby 1965, *No More the Tusker,* 183)

Few people appreciate the extent to which man-eating still governs the lives of people living in and around natural areas that harbor big cats. The victims are typically natives whose lives seldom intersect—emotionally or economically—with our lives in the industrialized world. Yet people in eastern and central Africa continually fall victim to big predators. Curiously, a communication medium of the twenty-first century—electronic publication on the Internet—has permitted a more immediate and continuous record of the predation that has shadowed humankind since our origins. Man-eating is probably not on the upswing, but reports of man-eating incidents and our ability to access them certainly are.

The range of lions has shrunk dramatically during the last century (discussed at length in Chapter 5), and what range remains is fragmented and isolated by areas that are dominated by people. This has served to bring lions and people into contact and conflict. Perhaps no better example exists than the lions of the Gir Forest in India, the last survivors of the lion's once-extensive Asian range. Once famed for their docility, Gir lions recently began attacking people, and these attacks were not directed at intruders in their territory but took place during forays from the Gir sanctuary. From January 1988 to April 1990, Indian biologist Vasant Saberwal documented eighty-one attacks resulting in sixteen deaths, nearly twice the number recorded over the preceding decade. Analysis showed that the attacks were clustered near high-density human populations—they were localized where, until

1987, lions had been fed to promote tourism. Accustomed to the presence of people, and with few options for acquiring territories of their own, such animals are more likely to enter conflicts with people.*

Man-eating and animal-human conflicts are the subjects of subsequent chapters. However, several recent events have brought the timelessness of man-eating—and the eternal specter it represents—very much to my attention. The first concerned a former Chicagoan, Wayne Hosek, who shot a man-eating lion in Zambia. The second involved a recent, scientific review of human-carnivore conflict over carcasses in Uganda, and the third concerned my wife, whose field party was recently attacked by a female lion.

The Man-eater of Mfuwe

On September 2, 1998, Wayne Hosek of Los Angeles, California, gave to the Field Museum the fully mounted specimen of a lion he had shot in Zambia (see Plate 3). In the flesh, the lion measured ten feet six inches in length and, despite being an adult male, was entirely maneless. In addition to various conversations with Mr. Hosek, the following account is drawn from three published sources: Walt Prothero's narrative of the hunt itself for *Safari* magazine (1996), Rob Vosper's account in the Field Museum's membership magazine, *In the Field* (1998), and a technical article written by two biologists, Koji Yamazaki and Teddy Bwalya, who were finishing fieldwork at Mfuwe when the incidents took place (1999). Since I wrote this chapter, Wayne Hosek has written a book-length account of his adventures and experiences, which is undoubtedly more detailed than any existing report; that work should be regarded as the definitive authority.

Mfuwe is a small town in the Luangwa River valley in eastern Zambia. It abuts two wildlife reserves, South Luangwa National Park (3,500 square miles) and the Lupande Game Management Area (1,870 square miles). Unlike the surrounding highlands, the Luangwa valley supports warm dry woodlands dominated by mopane trees and shrubs that grow in clay soils below 3,300 feet elevation. Baobabs are characteristic of this vegetative formation, as are seasonal pans. Mopane woodlands extend along the Luangwa and Zambezi river valleys in the east, stretch across northern Botswana to the Okavango Delta in the west, and blanket the Botswana-Zimbabwe frontier into Mozambique and the adjacent South African reserves, Kruger National

*Peter Jackson, Cat Specialist Group, Species Survival Commission, IUCN—The World Conservation Union; http://lynx.uio.no/catfolk/sp-accts.html

Park and Klaserie-Timbavati Nature Reserve. The Endangered Wildlife Trust characterized the region as one of high diversity but little endemism.

In 1990, the region surrounding Mfuwe was home to 9,871 people. Trophy seekers frequently hunted the area, and adult male lions were principal hunting targets. Scientists studying the region's lions concluded that continual harvesting of adult male lions would lead to disruptions of lion social systems (Yamazaki 1996).

The first three lion attacks were described with clinical clarity by scientific observers Yamazaki & Bwalya (1999). The first attack took place on 8 July 1991, as two boys walked at night along a road. About half a mile from town, a lion seized one and the other fled. When game rangers returned to the scene, they found only a piece of skull and some of the boy's clothes, alongside the tracks of three lions. Although baits were immediately deployed to lure them back within range, the lions did not return to the scene.

The second attack targeted a mentally impaired woman living at the edge of a village. On 28 August 1991, a visitor found the door to her hut broken open and bloodstained sand on its threshold; a drag mark was clearly visible on the ground. Her remains (the head, an arm, and parts of both legs) were found 150 yards away in tall grassland. A baited gun-trap placed near the spot where her remains were recovered attracted a lion but did not kill it.

The third attack took place a night later, as a boy who lived in Kakumbi set out after nightfall to visit a friend. He was attacked behind the ranger's office, which at the time contained numerous game scouts. One scout fired a warning shot at what he took to be a lioness, which was struggling with the boy; he missed, and the lion ran off. The boy had been bitten on the back of the skull, and his viscera were exposed by the attack. Despite being immediately hospitalized, he died the next day.

On 30 August 1991, lions were reportedly moving along the road from Mfuwe Bridge toward Kakumbi. Researchers and game scouts identified them as the L-pride, two adult females and a two- or three-year-old male. Both the older female and the young male had scars on their necks from wire snares set by poachers after bushmeat. A fusillade of shots killed the older female and wounded the male. Both younger cats escaped, but no further attacks were recorded in the Kakumbi area. At this point, though, the attacks moved to Mfuwe.

Details of the lion's next two victims are lacking, but just as Hosek boarded a plane in Los Angeles for his hunting safari in Zambia, the lion claimed its sixth documented victim. This was the subject of Prothero's colorful account. In Ngozo, a small hamlet a short distance from Mfuwe, an elderly woman had been sitting in the doorway of her mud-and-wattle hut, drinking millet beer from a gourd at sunset. Although her neighbors had already barricaded themselves behind latched doors, Jesleen remained outside her house, taking the cooling breeze of a hot September day. She didn't see the large male lion in the acacia thicket next to town and could do nothing to avert his charge.

When government scouts investigated the next day, they found blood-stained sand on the doorstep of her house and seven-inch pugmarks that could only indicate a very large lion. In the weeks since the first victim was claimed, several lionesses had been mistakenly shot in an attempt to end the depredations; no one suspected the large maneless male lion that sometimes associated with them. But after the departure of the game scouts, the male lion returned to the woman's hut in broad daylight, pushed the door ajar, and seized a gunnysack containing the woman's laundry. Carrying the bag to the center of the small village, the lion dropped the bag, stood over it, and began to roar. Alarming at any time, the lion's fearless domination amplified human fears and vulnerability and identified him as a sorcerer in the minds of village elders. The lion continued to carry the clothes bag about for several days, playing with it as a cat does with a dead mouse. Each day, villagers would encounter the bag at a different spot along the river beside town. Powerless and demoralized, people sealed themselves in their houses or traveled in groups.

Wayne Hosek arrived in Mfuwe via Lusaka, expecting to take a typical hunting safari. Zambia uses sport hunting to fund its conservation efforts, and hunters from around the world pay top dollar for hunting licenses and use fees. In turn, those funds are used to pay game scouts who regulate hunting, stop poaching, and compensate people for the inevitable offenses that wildlife commit when they trespass outside reserves. Hosek had permission to hunt various big game species, including lion.* After his professional hunter, Charl Beukes, explained the recent terror in the villages, Hosek agreed to use his lion permit to try to take the town's scourge.

Initially, they sat overnight in a blind 60 yards from a hippo hindquarter tied to a tree, but the inky blackness of the Zambian night foiled their efforts. In the morning they found lion tracks circling both the bait and their blind; they were alarmed by the big lion's stealth and care. His natural wariness

*Individual buffalo, kudu, bushbuck, grysbok, hippo, spotted hyena, impala, warthog, wildebeest, and zebra were also taken on this permit (Field Museum archives; U.S. Fish & Wildlife Service 3-177, dated 3 March 1992).

had only been enhanced as the females were shot in misplaced control efforts during the previous weeks. As the nights wore on, tormented by mosquitoes despite the nightly shellacking with repellent, dazed by jet lag, feverish with his reactions to inoculations, Hosek was on an emotional roller coaster. His vulnerability to attack while on the hunt left him breathless and with his heart racing, just as conversations with besieged townspeople filled him with indignation and anger at the pain and sorrow the cat had caused. There can be no better gauge of these emotions, and other details of the hunt, than Hosek's forthcoming book on the subject.

Finally, after twenty days of sitting in the blind, Hosek had his chance: in response to a new bait, the lion appeared at dusk, and a single 300-grain slug from his .375 rifle hit the lion as he was trotting toward the bait. Beukes also fired his .458 at the cat, grazing its foot before it disappeared. Some tense waiting ensued—no one who has watched a cat crouched patiently in ambush would mistake the absence of sound for the absence of danger. Certainly a sally to investigate the cat's condition could have had fatal consequences. A long time afterward, they heard a gurgling moan from the bush, and Beukes pronounced the cat dead.

As the group approached the dead lion, the lead tracker broke into the Kunda lion song. The townspeople poured out, jubilant, delivered—they slapped backs, shook hands, chanted, and wept. Although there are no words for the gratitude these people felt toward Hosek, one woman took his hand and looked deeply into his eyes, saying simply "*Zikomo kwambili*" (Thank you very, *very* much). A mild-mannered and soft-spoken man, Hosek had transformed her life without even speaking.

Although a resident of southern California, Hosek grew up in the Chicago area and still has family there. In fact, the "Man-eaters of Tsavo" and other dioramas in the Field Museum's exhibit halls—many mounted by the incomparable Carl Akeley—had whetted his initial interest in African mammals. At a press conference held at the Field Museum to commemorate his gift of the lion, Hosek withdrew from his pocket a small carved animal that his father had bought him nearly fifty years earlier on a museum visit. "Looking back to when I was a boy," he said, looking down at the talisman, "it is really surreal to think that one day I would be giving the Field Museum a man-eating lion" (Vosper 1998, 6).

Recent Carnivore Attacks in Uganda

Two University of Wisconsin researchers, Adrian Treves and Lisa Naughton-Treves, studied information in the Ugandan Game Department's archives for 1923 to 1994 on conflicts between large carnivores and humans (Treves & Naughton-Treves 1999). Many of these conflicts arise when humans attempt to appropriate the carcasses of carnivore kills, often leading to casualties. The records documented that wildlife encounters currently cause about 11 human casualties per year, 7 of them caused by large carnivores. Of the 393 attacks by large carnivores during the period they studied, lions accounted for 275 (70 percent) and were responsible for one-quarter of the injuries and three-quarters of the fatalities. Leopards accounted for most of the remainder. Men were attacked more frequently by lions, whereas women and children were more likely to be seized by leopards. Fatalities from carnivore attacks were mostly of women and children.

Physical anthropologists long believed that African cats had little effect on the course of human hunting and vice versa. However, the large number of interactions documented in the Uganda record calls this judgment into question. Louis Leakey is quoted as having said, "Man is not cat food," but the analysis of Treves and Naughton-Treves shows otherwise. Humans are occasionally consumed as a result of confrontations over carcasses. Significantly, only 14 percent of the attacks by lions and 15 percent of those by leopards involved disabled or wounded animals. In other words, these appear to be normal ecological interactions, not the result of disease or pathology. The majority of the attacks come from healthy animals behaving in a normal fashion, defending their kill from theft by humans as they would in encounters with other carnivores and scavengers.

Humans and carnivores probably compete less intensely today than they did in prehistoric times, but predation on humans today by lions and leopards differs only in number, not in kind, from what our Pleistocene ancestors experienced. Surprisingly, possession of weapons had little effect on the likelihood that these encounters would result in human deaths. At close range, spears are as effective against these predators as rifles. The balance between humans and large carnivores may have shifted a bit, as more people live in cities outside of danger from animal predators, but the stakes haven't changed a bit.

A Recent Episode from Uganda

My wife, Barbara Harney, is also a zoologist, specializing in behavioral and community ecology. In 1994, she spent ten weeks in Africa, teaching a field course to British student volunteers on how to survey and inventory small mammals. That collaborative program between the Field Museum and

Frontier, the Society for Environmental Exploration* (a volunteer organization based in the United Kingdom) aimed to inventory vertebrate species in all of Uganda's game reserves. Expanded lists might be used both to promote ecotourism and to serve as a baseline for conservation efforts in protected areas. Most of Uganda's reserves had been ravaged during the political upheavals of the 1970s and 1980s, and current inventories were lacking. The group faced numerous difficulties, most related to the logistics of fieldwork during the rainy season. I received a letter from Barbara about six weeks into her trip:

Dear Bruce,

I hardly know how to begin to tell you about what happened yesterday—I am not quite sure how I feel about it myself. I'd just finished skinning and prepping the day's catch (an uninspiring lot of Lophuromys, Aethomys, Praomys [all field mice]) and had sat down for tea at 4 pm. It had been nearly nonstop work since 6:30 am amid two torrential downpours. The reconnaissance team [that had] headed towards Ishasha on the Zairean border of the reserve drove up to camp and insisted that they had an interesting mammal for me to work up.

With little more than a nod I walked towards the truck with Christie Allan [research coordinator for Frontier Uganda] insistently beckoning me on. On the bed of the truck was a sight I'll never forget—a dead lioness—shot by Joseph Arinaitwe, the game guard, to save the life of one of our team, James McCaul. They had been marching across the savanna when three lions shot out from an acacia tree 15 meters from them. The game guard fired one of his two shells and successfully turned two away. The third continued, [and] Joseph shot it through the mouth after inquiring from James whether it was permissible to do so. The lion fell 5 meters from them.

I can't tell you how I felt staring at that dead beast in the truck, realizing that one of us might be dead but for its life. Nor did it help realizing the full burden of disposing of it properly was to fall on me as all eyes turned my way in question of what to do next. I can only tell you I knew I'd do my best, although I could scarcely contain the tears that welled up inside me or the shaking of my knees in considering what had happened and what needed to be done.

*http://www.frontierprojects.ac.uk

Three of us, James, Christie, and myself, set about skinning it under my direction; I panicked a bit when I could not disarticulate the claws, but perseverance with the scalpel finally paid off. She was magnificent, maybe 280–350 lbs (seven people lifted her into the truck), hugely muscled and nursing, with milk still in her nipples. We lost the light at 7:00 pm and continued working with lanterns and head lamps, with an audience of 15–20, everyone expecting me to know what to do and how to do it. After removing the skin we salted it and rolled it and carried the gutted carcass to Alfred's (game warden) storehouse for the night. This morning, Christie and I decapitated the carcass, fleshed it, removed the brain and soaked the skull in alcohol. We dug a pit four feet deep and buried the carcass and entrails some 15 feet from Alfred's house—next phase they'll dig it up and have a look at the skeleton. We put the skin on the rack to dry—I just rolled it up before the rain and it seems to be drying OK.

Shooting a lion is a pretty heavy scene here. The game guards and villagers celebrated "survival" with banana wine and singing. I just felt numb, drank two beers (we now have beer and soda from the local village sold in camp to supplement our food budget) and collapsed. As luck would have it, the batteries in my camera failed just as I tried to document all this but a couple of folks took plenty of shots while I was deeply into it, so all is not lost. Tomorrow, we'll drive to Kasese, then on to Mweya in QENP [Queen Elizabeth National Park] where we'll deliver the skin and skull to the large-mammal biologist, John Beebwe.

It may be an understatement to say that all else feels anticlimactic now, I guess I still feel we had no business walking willy-nilly in an area densely populated by lions (they had already seen 11 before the attack) where several villagers had been mauled previously. It was a sober reminder that … there are large mammals and snakes here that can kill you, but now I would seriously question the wisdom of sleeping 30 feet away from a pool filled with hippos while attending an all-night bat netting simply on [the] assurance that hippos in the water aren't dangerous, and they aren't likely to leave the water with humans about. Christie and I at least have agreed that returning to Ishasha to run transects is foolhardy given the lion threat and the scarcity of [rifle] shells—we have only six remaining.

I am truly disheartened that as a result of our activities there is one less lion in Africa (at least—given her hungry cubs at home). So I woke this morning and sobbed my heart out, feeling an emptiness and sadness that made me want to hold the girls [our dogs] and be assured of their corporal well-being. You know what an absolute mush I am about female mammals and their babies and here I am in the thick of it. It is an adventure and I believe I've performed well under the circumstances. But the lioness is enough—my African experience is assured. Take care and I'll see you soon.

Love, Barbara

Chapter 3
Killing Behavior and Man-eating Habits

Except when resting and in the breeding-season, the whole career of a lion may be summed up in the single word rapine. (T. Roosevelt and E. Heller 1914, *Life-Histories of African Game Animals*, 169–170).

rapine: n. The action or an act of seizing property etc. by force; plunder, pillage. (*The New Shorter Oxford English Dictionary*, 1993).

Lions usually kill their prey with a vise-like bite to the victim's throat, applied from the front. Most cats orient their attacks toward the neck, but smaller cats generally bite the nape or the back of the head. In this maneuver, the canines are sunk into the neck, causing death generally by severing the spinal column or penetrating the braincase. German cat specialist Paul Leyhausen saw the ubiquity of this maneuver among cats and related carnivores as evidence for its being the primitive or ancestral state of the killing bite—how proto-cats killed their prey (Leyhausen 1979).

But some prey animals, particularly cattle, have very thick necks that cannot be easily punctured in this manner, while others have horns or antlers that could damage a cat employing such an attack. Consequently, the great cats also employ a derivative attack, the "throttling grip," in which the throat (or sometimes the muzzle) is seized and the trachea or mouth clamped shut. Death in this case is by asphyxiation. The throttling grip involves the same general orientation toward prey and targets the same body region for attack. Certainly, smaller cats find it unnecessary to suffocate a mouse or rabbit, although that would also kill their prey. The throttling grip was probably the predator's counter in this predator-prey "arms race." Scottish Museum zoologist Andrew Kitchener devotes a chapter of his book to the diverse means by which cats kill and eat (1991).

The larger pantherines sometimes target the muzzle in a related method of attack. Here, the throttling grip is applied to the face, clamping the prey's mouth and nostrils closed and stifling respiration. This action seems easily derived from the throttling grip—the locus of the bite is merely shifted forward, with suffocation as the same ultimate consequence.

Lions and tigers also target the muzzle in a fundamentally different maneuver. They leap onto the victim's flanks while facing in the same direction and grasp its muzzle with the steely claws of one paw. Usually the teeth grip the nape as the cat's weight topples the prey, causing it to fall forward onto its neck. If the neck is broken, this deft maneuver virtually eliminates a potentially risky struggle with a large, powerful prey animal. Using it, lions can feed on a wide variety of formidable animals without major risk.* For example, Teddy Roosevelt and Edmund Heller observed lion bites on the napes of both hartebeest and zebra during their famous East African safari (1914). These probably represented stabilization bites offering the lions purchase in a "topple attack" rather than the killing bite used by smaller cats to sever the spinal columns of their prey.

The "topple" depends on the cat's unparalleled sense of balance and on its immensely powerful arms and shoulders. Noted taxidermist and sculptor Carl Akeley offered this tribute in his book *In Brightest Africa:* "... perhaps the most impressive thing about a lion is his foreleg. The more you know about elephants the more you regard the elephant's trunk. The more you know of lions, the more you respect the lion's foreleg and the great padded and clawed weapon at the end of it. It is perhaps the best token of the animal's strength. It is probably two or three times as powerful in proportion to weight as the arm of a man. He can kill a man with one blow of his paw" (1920, 76). Both lions and tigers are known to kill domestic stock precisely so, with a single decisive blow from the forepaw.

There is often real drama in cat kills, and frequently the predator requires some latency period in which to recover. If the kill has been made in the open, and the day is dawning, lions will avail themselves of nearby shade. Akeley repeatedly witnessed lions transporting zebras a distance of fifty yards. At six hundred to eight hundred pounds, zebras weigh about

*Young and clumsy or old and weak lions commonly suffer injury in attacks on larger prey, and these can have important implications for animal-human conflicts, including man-eating. But only buffalo constitute an ever-present danger for lions of all ages, including those in their prime (Roosevelt & Heller 1914).

twice as much as lions do. To move a zebra, the lion straddles the zebra, seizes it by its neck, and drags it forward a few steps, stopping to rest frequently. If the lion fears disturbance, the prey may be dragged a considerable distance to suitable shelter, and this is often the case when lions take human prey or livestock. It also occurs in situations where the carcass may be stolen, either by hyenas or by human scavengers. In cases of potential conflict, lions often feed quickly and rest far from the carcass.

Sometimes, lions transport prey much greater distances. In 1977 in northern Kenya, Smithsonian paleobiologist Kay Behrensmeyer witnessed a lion with a newly killed crocodile seven feet long in some shade plants about a third of a mile from Lake Turkana. After the lion ate and abandoned the carcass, she approached it to note damage to its skeleton. She found the crocodile's jaws were clamped tightly around a toothbrush tree (*Salvadora* sp.), signifying that the crocodile had still been alive when it reached the spot.

Often the victim is licked all over for some moments before the meal begins. This activity is usually initiated by blood flowing from freshly made wounds. Animals killed by attacks involving holds to the throat often have severed carotid arteries or jugular veins that create large pools of blood, which are bitten and drained at the onset, especially in arid areas. Licking can be done with such steady purpose that the hair is removed in patches and the skin abraded and stripped by the steady rasping licks. The historic man-eaters left behind human remains that appeared "sucked dry" of blood, no doubt a product of this activity.

Lions develop close, lifelong bonds with pride-mates and show various signs of affection and cooperation, from grooming each other to nursing each other's young. But all bets are off at a kill. Lions lack the rigid dominance hierarchy evident in other animal societies, so that it is every cat for itself, until the larger or more motivated animals prevail in highly competitive situations. "Each lion in a group knows and responds to the fighting potential of every other member. It is a system based on the amount of damage each animal can inflict on an aggressor; it is a system based on a balance of power...." (Schaller 1972, 136).

The body is typically bitten open at the flank or belly, and diners first focus on the viscera. The stomach is promptly removed intact, its undigested contents are stripped, and then the organ is eaten. The lungs, liver, and kidneys all follow in short order, but the intestines are usually removed some distance from the carcass and often partially buried under grass or earth. After these appetizers, lions usually begin consuming the soft flesh of the hind legs and rear, eating their way toward the head and shoulders (Astley Maberly 1963; Guggisberg 1975). Skin and hair are devoured with the meat, so the droppings of lions are usually full of hair (Astley Maberly 1963). This habit enables scat analysis to reveal the full range of lion diets.* Given the opportunity, lions will eat twenty to sixty-five pounds at a sitting.

Any experienced African tourist knows that waterholes are favorite resorts for lions. Lions frequently rest near waterholes and other drinking places where animals converge to slake their thirst. Here they minimize trips for drinking and may be able to ambush prey (Fitzsimons 1919). Acclaimed naturalist and author Charles A. W. Guggisberg was impressed by the frequent recourse that East African lions have to water, drinking daily or every two or three days and rarely straying more than six to ten miles from permanent water (1975). However, in this respect as in others lions are adaptable and variable. In Somalia, lions living in the Haud only rarely drink from pools of rainwater (Swayne 1895). More commonly, the water they need is obtained from the blood and viscera of their prey.

In parts of their range, lions supplement water obtained from animals with vegetable sources. Kalahari lions seem to rely on *tsama* melons (*Citrullus lanatus*) for this purpose, as do Kalahari Bushmen, jackals, and various game animals. One pride that South African zoologist Fritz Eloff tracked continuously did not drink water for at least nine days; another—one with young cubs—drank more regularly (1973). In that study, individuals were seen more than 70 miles from open water during a summer drought, with daytime temperatures soaring above 106° F. This represents at least a two-day trip without water, each way. The combined stresses of temperature and water may drive Sahara lions underground. During daytime hours, lion survival in regions where temperatures soar above 110° F probably depends on being able to avoid lethal heat "loads" by sheltering in burrows or caves (Guggisberg 1975).

Man-eating Is Habit-forming

Most lions never threaten a human during the course of their lives, and it would be erroneous and irresponsible to discuss man-eating incidents without underscoring this simple fact. Examples abound of lions' avoiding con-

*Direct observations of lion meals are inevitably biased with respect to prey species. A large group of lions may camp for days on a buffalo carcass, attracting a host of would-be scavengers that make locating these kills very easy. Conversely, it is very difficult to witness or locate a lone lion feeding on a hyrax or dik-dik, as the meal is completely consumed before the record can be made.

tact with humans wherever and however possible. One of my favorite quotes, mainly for its ridiculous and anachronistic subjectivity, is the following. In his report on the first zoological expedition to Africa sponsored by a North American museum, the Field Museum's 1896 expedition to British Somaliland, Daniel Giraud Elliot said, "Judging from our experience with them, [lions] are most cowardly in disposition, and avoid man's presence whenever possible. Of course, if wounded and surrounded so that escape seems impossible, the Somali lion will show fight, as any other animal will, even a rat, but his principal idea seems to be when followed to put as much ground between himself and his pursuers as possible" (1897, 144).

Martin Johnson, who spent years in the bush in East Africa, said, "It is only fair to add, however, that even at this time I was not afraid of being eaten by a lion. The belief that a lion is a man-eater is generally incorrect. Lions enjoy zebra and giraffe meat best of all; human flesh does not generally appeal to them. In my six years' residence in Africa in lion country I have never been able personally to trace an authentic case of a man-eating lion. Often I have heard of them; but, when I run the facts to earth they always turn out to be nothing more than wild rumor" (1928, 263). While his statement is generally true, and his perspective useful, man-eating lions are well documented, and the ensuing discussion focuses on these.

C. A. W. Guggisberg identified a panel from near Nimrud, Iraq, now in the British Museum, as one of the earliest representations of a man-eater. Made in the eighth century BC in Assyria, it shows a lion seizing a human by the throat (Guggisberg 1975). Throughout history, this large, adaptable cat has taken advantage of opportunities when human beings, and sometimes very little else, are available. Because they are the dominant predator wherever and whenever they occur, lions are logical objects of fear and respect, of symbolism and metaphor, of news and legends. But when they turn to man-eating, lions are something else.

In lions, man-eating very quickly becomes a habit, a routine, a way of life. As mentioned in Chapter 1, when Patterson was pursuing the Tsavo man-eaters, he repeatedly laced the carcasses of transport animals that had died from sleeping sickness with strychnine, placing them where the lions would likely encounter them. "But the wily man-eaters would not touch them, and much preferred live men to dead donkeys" (1907, 106). George Rushby encountered quite a few man-eaters as a professional ivory hunter in Mozambique and adjacent territories. He said: "Lions, when they turn man-eater, make human flesh their staple diet and will only kill and eat game or domestic stock when driven by extreme hunger. The killing and eating of humans by man-eating leopards is more spasmodic and in some cases they stop killing humans for lengthy periods." (1965, 129). Rushby attributed this difference to the natural diets of these species. Lions seldom eat primates of any kind, the occasional baboon or vervet excepted, while leopards frequently depend on them. He presumed that humans would taste similar to their primate relatives.

Taste preferences are also apparent in John Hunter's interpretations of man-eating. He believed that man-eating sometimes arose by accident, as when a lion attacks a herd of cattle. He said, "In trying to get at the cattle, the lion may kill a herdsmen, just as he would knock down any obstruction that stood in his way. If for some reason the cattle then managed to escape, the lion may return to his kill and begin feeding. This combination of circumstances is rare, but when it occurs the lion almost invariably becomes a man-eater. However, once a lion definitely acquires a liking for human meat, he will go to the most amazing length to satisfy his craving. I have even known confirmed man-eaters to charge through a herd of cattle to get at the herdsmen" (1952, 205–206). Although the authority of both statements is indisputable, it is difficult for most of us to subscribe to the view that it is taste per se that determines the behavior of these carnivores. But maybe humans are special when it comes to taste. Rushby interrogated an old cannibal in the Republic of the Congo who still practiced cannibalism. "I asked this old cannibal why he preferred human meat to that of game. He replied, 'There is no comparison at all.... It tastes better and is much sweeter than elephant meat or any other meat including gorilla and chimpanzee'" (1965, 129–130). Man-eating supposedly becomes so ingrained in lions that cattle were perfectly safe in the country hunted by the infamous Njombe lions (see Chapter 4): "On one occasion they took a herd boy off the back of a cow on which he was riding without harming the cow in any way" (1965, 193).

Man-eating lions are surely persistent. One of the most gripping episodes comes from Southern Africa in the nineteenth century and is related by Frederick Fitzsimons: "Gordon Cummings tells a harrowing tale of a hungry lion that prowled around his camp one night, roaring in marrow-freezing manner, intent upon getting at his oxen. One of his Hottentot drivers got up from the fireside to incite the dogs to rush out and scare the 'King of the Forest,' and to cast a few firebrands in his direction, as lions usually have a great dread of fire. The Hottentot, immediately upon lying down again by the fireside, was pounced upon by a huge, shaggy-maned lion which rushed out from the inky darkness, seized him by the back of the head, and in spite of being belabored by the victim's comrade with a firebrand, the lion carried his quarry away into the bush. For some hours

Cummings and his men sat listening to the cracking and crunching of the Hottentot's bones, accompanied by the lion's occasional growls of satisfaction.... At early dawn they ventured out, and all they found of the Hottentot was some ragged pieces of clothing, a few mangled remains of flesh, and a boot with a foot in it" (1919, 107). Col. Patterson claimed, "Having once marked down a victim, they [the Tsavo man-eaters] would allow nothing to deter them from securing him, whether he were protected by a thick fence, or inside a closed tent, or sitting around a brightly burning fire. Shots, shouting and firebrands they likewise held in derision" (1907, 28). Roosevelt also noted that if his first attack is unsuccessful, a man-eater often perseveres (Roosevelt & Heller 1914). The dauntless lion that attacked O'Hara near Voi certainly epitomized this tendency.

Stalking Human Prey

Man-eaters seem to become by turns brazen and wary. As a man-eater continues to claim victims, through trial and error it quickly begins to learn, understand, and even anticipate human behavior. Just as a juvenile lion learns the drinking schedules of impalas, the grazing habits of warthogs, or the bedding preferences of old solitary buffalo bulls, so a man-eater assembles a profile on humans. Where humans still live in wilderness areas, their homes, fields, neighbors, and communities all lie within the hunting patrols of lions. Despite vigilance and measures of avoidance and defense, a vulnerable person is bound to run across a hungry lion, creating the opportunity for an attack.

And the discovery that people are so easy to kill, or perhaps so good to eat, is almost certain to initiate a new cycle of man-eating. Rushby said, "When any such outbreak occurred it was advisable to place as many Game Scouts as possible in the area and concentrate all efforts on the problem.... If a man-eater continues to kill and eat people for any length of time it develops an almost supernatural cunning. This often makes the hunting down and killing of such a lion a lengthy and difficult task." (1965, 183).

It is unclear whether lions differ significantly in this respect from tigers or whether the different human tolls claimed by these cats derive from cultural or geographic differences. John Hunter certainly thought so: "I have read with great interest several excellent accounts of hunting man-eating tigers in India. I must say that I am very much astonished at how the business is handled in that country. Apparently after a tiger has killed and eaten four or five hundred people, some young subaltern, tourist, or local sportsman decides to have a go at the animal. In Kenya, a man-eater is regarded as a dangerous menace that must be killed immediately. There is no element of sport in the hunt. As soon as a man-eater is reported, the game department instantly gives the animal top priority above every other consideration" (1952, 205–206). Any delays simply permit the lion to refine its hunting skills further.

Game control officers responsible for killing man-eaters use any means at their disposal to accomplish this: traps, poison, gun sets, and *machans*. "If a man-eater kills a native and leaves part of his victim uneaten, he will almost certainly return to the body just as he does in the case of wild game. If the hunter poisons the body, he is virtually certain of getting the lion" (Hunter 1952, 206). Hunter recounted the curious story of his friend Tom Salmon, who was a ranger in the district surrounding Makindu, Kenya. A lion had killed the mother of a local Wakamba chief. Following the lion's trail in daylight, Salmon and the chief found first her arm and then her partially eaten body lying in an open area without nearby trees or bush. Unable to fashion a *machan*, Salmon asked for permission to use the remains as bait. The chief consented, her body was laced with strychnine, and the lion was found dead atop it the next morning. However, not all man-eaters behave in this manner. In Mozambique, John Taylor found that man-eating lions seldom returned to their kills, complicating the process of dispatching them.

Initially, most man-eaters display normal hunting and killing behavior. All lions must have such skills to survive their first years, and these skills are continually honed by practice and refinement. Most attacks of man-eaters come at night, when lions typically forage, and rely on the element of surprise, a normal hunting strategy (Roosevelt & Heller 1914). Because lions are accustomed to preying on large, dangerous species, they naturally minimize their own risk by first identifying the vulnerable and then concentrating their attacks on that individual. The same strategy is in effect with man-eaters. Sir Richard Burton, writing in Somalia in 1854, said, "The people have a superstition that the king of beasts will not attack a single traveller, because such a person, they say, slew the mother of all the lions . . . he is a timid animal, much less feared by the people than the angry and agile leopard. Unable to run with rapidity when pressed by hunger, he pursues a party of travellers as stealthily as a cat, and arrived within distance, springs, strikes down the hindermost, and carries him away to the bush" (Waterfield 1966, 143). Burton's observation still holds, as John Taylor testified: "All maneaters, whatever their type, will invariably take that almost inevitable straggler—someone who has stopped for a moment to adjust his load, to tie a sandal, or to relieve nature" (1959, 56).

Man-eating lions operate by exploiting momentary lapses in vigilance. On his first trip to Africa, Akeley learned of the experience of an English traveler and his wife in Somalia: "They were intent on getting a lion by 'baiting'—that is, they killed an animal and left it as bait for the lions while they hid in a thorn *boma* which they built near by. There was only a small hole in the *boma* through which to watch and shoot. They stationed a black boy at this hole to watch while they slept. They awoke to find that a lion had stuck his head into the hole and killed the black boy—bitten his head clear off, so the local story goes. However, no one knows why the lion killed the boy in this case for, of the three possible witnesses, two were asleep and the third dead" (Akeley 1920, 60–61).

Recently, another lapse in vigilance led to death, in August 1999, when an eighteen-year-old Briton was killed while camping in Zimbabwe by at least a dozen lions without prior history of conflicts with people. The attack was led by an elderly lioness and took place at 1:30 am in Matusadona National Park, 240 miles northwest of Harare. The young man had failed to secure his canvas tent flap, which permitted at least one of the lions to enter and initiate the attack. The man fled outside through camp, where he was surrounded by the rest of the pride. Despite efforts to repel the lions with flares and gunfire (two animals were killed), police later retrieved only the man's head, ribs, liver, and heart from the scene (Anonymous 1999).*

Man-eating Is a Learned Behavior

Like any other hunting and killing behavior, man-eating must be practiced and learned. Most investigations of man-eating seek to identify the initial trigger for man-eating behavior, its original cause. Yet once acquired, this behavior is transmittable, through imitation and cooperation. Mother tigers sometimes begin hunting humans to meet the increasing energy demands of their offspring. It is only natural that cubs brought up on human flesh and either witnessing or assisting in the dispatch of human prey will come to regard people as targets of predatory behavior (Astley Maberly 1963; Hunter 1952). The transmission of man-eating habits from individual to individual may lead to regional outbreaks of man-eating as cubs raised as man-eaters multiply.

Other members of social groups (pride mates and members of male coalitions; see Chapter 6) may also acquire man-eating habits by imitation and cooperation. Roosevelt and Heller observed: "Doubtless the lion, like

*http://www.igorilla.com/gorilla/animal/1999/lions_kill_boy_in_zimbabwe-inquest.html

other animals, varies in character and habits from place to place; and if by any chance a single lion in some particular locality learns how to prey on an animal not ordinarily attacked, other lions may readily learn to follow his example. At any rate, it sometimes happens that lions in one district as compared to another have entirely different customs as to what game they prey on" (1914, 181).

In many rural areas, the late afternoon and early evening are a time of heightened vulnerability to lions. Weary people are returning home from the fields, children are playing in small groups, women are cooking outside their huts, lovers are meeting in discreet trysts, or neighbors are simply exchanging gossip. As the shadows lengthen, people have shrinking perceptual powers, while the adaptations of lions make them increasingly effective and deadly—lions are actually most effective when hunting on moonless nights (Van Orsdol 1984). To enter a village, man-eaters will often use thickets, those growing alongside water sources or maintained nearby for cooking fuel. Once in the midst of town or close by, the lions have only to wait for a moment of vulnerability—a pounce or a charge and tackle are all that are required to claim freestanding victims. Sometimes a victim is able to take shelter in a mud-and-wattle house, which is then besieged or broken into.

All cats are naturally drawn to the neck, and most man-eaters are no exception. And it does not take much to kill an unarmed person taken by surprise. Any of the three primary methods used by lions to kill normal prey may be used to dispatch humans: the nape bite, the throttling grip, or a powerful blow with the foreleg. The thin skin of humans and the delicacy of the human neck and skull mean that few humans survive a man-eater's attack; when they do, it is mostly owing to a rushed or fumbled attack and a misdirected bite.

Thirty years ago in Vietnam, a tiger attacked members of a recon team camped about six miles east of the Laotian border, in an incident recorded by Sgt. Bob Morris. Bad weather had delayed the team's evacuation from the site, and they had posted two-man watches and begun taking turns sleeping. The tiger struck with stealth, seizing one of the men in its mouth, who immediately began to scream. PFC Roy Regan, just feet away from the victim, saw the tiger with the man in his mouth. "All I could think about was to get the tiger away from him. I jumped at the tiger and the cat jerked his head and jumped into a bomb crater 10 meters away, still holding his prey." The Marines immediately followed the tiger and killed it with their M-16s. The injured Marine suffered lacerations and bites on the neck. The same animal is thought to have killed a Marine earlier in the same vicinity.* Although pumas cannot carry victims in the same manner as the great cats, the vast majority of their attacks involve leaps that knock a victim down and bites to the neck and adjacent head and shoulders.

More recently, on 5 April 2000 in Recife, Brazil, five lions ate a six-year-old boy who had approached their circus cage too closely. One lion struck, pulling the boy through the bars, clamped its jaws on his head, and shook him violently as the other lions attacked. A circus employee told a local television station that the lions had not been fed for a week.† This was an attack by captive-bred lions that had never hunted for themselves, presumably involving innate behavioral elements. Many wild man-eaters, including the Tsavo lions, the "Kima Killer," and the Mandimba man-eater (described by John Taylor, 1959), also seized adult men by the head or face, sometimes killing them instantly .

Those surviving the onslaught of a lion's attack report various sensations. The famous missionary Dr. David Livingstone experienced no sensation of pain when he was bitten by a lion, and attributed this to the mercies of Providence. But acclaimed explorer and naturalist Frederick Selous strongly disagreed. He noted that a Boer settler who was bitten by a lion in the arm, hand, and thigh had experienced "the most acute anguish" each time the lion bit him, "so that I can but conclude that this special mercy is one which Providence does not extend beyond ministers of the Gospel" (Selous 1881). Undoubtedly, shock responses have much to do with perceived pain.

Although man-eaters often begin their careers with minor, temporary modifications of normal hunting and killing behavior, they often develop abnormal behavior patterns as they devote themselves more exclusively to human prey (Taylor 1959). Taylor found this true of man-eating lions, tigers, and leopards. He noted that normal lions, especially females, will routinely charge a hunter that fires a weapon at them—not invariably, but usually. However, he could not "recall a single maneater or his mate that charged me before being wounded" (1959, 24). This reluctance to confront or challenge humans in retaliation or anger is at distinct odds with attacks motivated by hunger. Such complex reactions highlight the ambiguity behind "belligerence" and "aggression" when used to describe interactions between lions and people. Because of the danger they pose to us, any actions by such predators—even normal, purely defensive ones—can be construed as aggressive. All predators appear aggressive to their prey.

*http://grunt.space.swri.edu/tiger3rd.htm

†from the Electronic Telegraph, London; http://www.igorilla.com/gorilla/animal/2000/Lions_eat_boy_in_brazil.html

Although lions rarely move natural prey more than a few yards, usually for shade, prey transport is typical behavior for a man-eater or marauder. "When he [the lion] is the hunter he always retires with his victim, as soon as he has caught it, out of reach of vengeance, although he may only go for a distance of a few hundred yards, being confident in the shelter yielded by a dark night. This is entirely unlike the lion's conduct with other prey; if a zebra or hartebeest is killed, the lion stays on the spot with his victim, and may eat it where it has fallen or drag it a few yards to a more convenient spot" (Roosevelt & Heller 1914, 178–179).

Rushby concluded that the infamous man-eaters of Njombe, in southern Tanzania, actually used relays, the lions taking turns carrying their corpses as much as a mile from the villages where they claimed their victims (1965). The Njombe lions also differed from normal lions in their ranging behavior. Whereas lions typically hunt and travel in the evening or at night and are sedentary by day, the man-eaters generally made their kills in the afternoon or evening, spent the night feeding and resting, and moved away before dawn. Often the next night found them 15 to 20 miles from their previous bivouac. "They seem to know that the survivors of the attack will redouble their precautions so that it will be extremely difficult for them to bag another just then" (Taylor 1959, 46).

Fear and Superstitions About Lions

Superstition is another factor that complicates attempts to control man-eaters; in fact, superstitious beliefs may even permit outbreaks to spread. As a man-eater (or group of them) claims victims, and the human toll begins to mount, the freedom, joy, and spontaneity of villagers is oppressed. People develop a sense of resignation, with the inexorable loss of friends, family, and neighbors. "Superstition and demoralisation play a very considerable part in preventing the villagers from taking any concerted, planned action against their adversary. Roads are deserted, village traffic comes to a stop, forest operations, wood-cutting and cattle-grazing cease completely, fields are left uncultivated, and sometimes whole villages are abandoned for safer areas. The greatest difficulty experienced in attempting to shoot such animals is the extraordinary lack of co-operation evinced by the surrounding villagers, actuated as they are by a superstitious fear of retribution by the man-eater, whom they believe will mysteriously come to learn of the part they have attempted to play against it" (Anderson 1955, 8–9). Although Anderson's observations applied specifically to tigers in India, superstition typically plays some role in man-eating attacks by lions and leopards in

Africa. A resort to the supernatural appears to follow quite naturally from the powerlessness people feel toward man-eaters.

After the first shocking attacks at Tsavo, superstitious workers began to regard the man-eaters as supernatural. Patterson observed, "They were quite convinced that the angry spirits of two departed native chiefs had taken this form to protest against a railway being made through their country, and by stopping its progress to avenge the insult thus shown to them" (1907, 21). Many years later and a country away, Rushby encountered difficulty even in gathering basic information about lion attacks in Tanzania: "They [local people] stated that these lions were not real lions at all but werlions [*sic*] or *simba mtu*. That is to say, certain unknown men, possibly one or two of them returning from the dead, took on the appearance of lions. After they had killed and eaten some person or persons they would travel a distance as lions and then revert to their normal appearance as men" (1965, 187). This is ample discouragement for informants to step forward with vital information.

Rushby also described how a witch doctor named Matamula used the lions' Reign of Terror to extort tribute and reclaim power. Matamula had been headman of Iyayi village until shortly before the lions began killing people in 1932. He was removed from this position for corruption but aspired to regain it. Matamula claimed that the lions were under his complete control, managed for him by two assistants. One of these supposedly managed the lions within the district but at a safe distance of some 30 miles from Iyayi, while the other delivered groups of lions into selected areas to kill and eat certain individuals, at Matamula's command. To obtain protection from him, local Africans paid tribute: livestock, money, or labor. As Rushby took one after another of the Njombe man-eaters, the shrewd Matamula realized his days were numbered and agreed to suspend man-eating altogether if he were reinstated as headman, inciting a popular revolt against his worthy and honest successor. By pure coincidence, Matamula reclaimed his office a month before the last of the man-eaters was killed, cementing the superstition that had carried him back into office (Rushby 1965).

Superstition continues to charge human reactions to man-eating. A German biologist working in Tanzania, Marcus Borner, reported that over a two-year period more than 250 lions had been killed in southern Tanzania in connection with human deaths. Sadly, it appears that many of these were not man-eaters. In Tunduru, where the attacks took place, some of the deaths were actually paid murders committed by so-called lion men (Anonymous 1989). Manipulation of human fears of lions is described in more detail in Chapter 4.

Chapter 4
Why Do Lions Kill People?

Many factors contribute to man-eating behavior in lions—from old age and infirmity, to the shortage of alternative prey, to making the natural transition from scavenging on human corpses to attacking live humans. This chapter reviews these causes and others, to better understand the man-eating attacks of lions.

The Primary Cause of Man-Eating: Opportunity

Where lions are much hunted it is doubtless true that they grow so wary of man that only the dire want produced by utter feebleness could make them think of preying on him; but where they are less molested, their natural ferocity and boldness make it always possible that under favorable circumstances a hungry lion, not hitherto a man-eater, will be tempted to kill and devour a man, and will then take to man-eating as a steady pursuit. Many noted man-eaters—those killed by Mr. Patterson, for instance—have been full-grown male lions in the prime of life and vigor." (T. Roosevelt & E. Heller 1914, *Life-Histories of African Game Animals,* 178)

Opportunity is surely a primary cause of man-eating. We tend to overlook this obvious fact because opportunity is so easily overshadowed by capability. The predatory abilities of carnivores have been honed by millions of years of natural selection. The structure of teeth, the powerful design of jaws, sinewy limbs, and sharp claws—all optimized by evolution for mayhem—command our focus and respect. Yet invariably associated with that fearful equipment is a malleable and adaptable set of behaviors that allows these animals to tailor their actions to achieve maximum benefits. And it is the adaptability of carnivores as much as their fearful adaptations for meat-eating that make them so dangerous to people.

Frequency of Encounters

There are numerous indications that wild carnivores have adapted to living in a human-dominated world. Several years ago, the *Chicago Tribune* carried reports of coyotes living in Lincoln Park, a fashionable lakeside neighborhood in the heart of metropolitan Chicago. The metropolitan region comprises a highly developed urban and suburban mosaic covering a radius of some fifty miles. The coyotes, which have only lately invaded this part of Illinois, were feeding on ducks that had grown complacent owing to the bustle of pedestrians. Of course, carnivores must increasingly retrofit their habits and behaviors to accommodate the inexorable expansion of humans into natural areas. Sadly, many are unable to do so rapidly enough.

This continuing encroachment of people into wilderness areas reduces wilderness (where carnivores can live naturally without human interference) and increases "edge" (where humans and carnivores come into contact). Often this contact has tragic consequences. Some of the clearest examples of opportunism by carnivores are provided by more remote settlements in California. Over the last century (1890–1990), fifty-three attacks by pumas on humans were recorded in the U.S. and Canada combined. Nine of those attacks resulted in ten human deaths. But today, on average, there are three to four attacks annually by pumas on people in North America, a rate six to eight times the historic average.*

Hunting is an immensely complicated procedure, and all mammalian predators must learn it, either by imitation or by trial and error. No carnivore could afford to have such complex responses genetically "hardwired," via innate, instinctive behavior patterns. In the face of a fixed predation threat, a simple shift in the behavior of prey populations would make them winners in the incessant struggle between predators and prey that has been likened to an "arms race." Consequently, each individual predator must learn the lessons anew. This fact explains the prevailing trend for predators to be smarter than their prey; certainly, cats and dogs outperform rabbits and guinea pigs in various sorts of intelligence tests. A key advantage to using intelligence and malleable behavior as a strategy for adaptation is that

*http://tchester.org/sgm/lists/lion_attacks_nonca.html; http://www.igorilla.com/gorilla/animal/2001/cougar_kills_skier.html

it tailors each individual's hunting patterns and diet choices to its immediate surroundings.

Young, inexperienced carnivores are often unsuccessful in hunting, and they have not yet had experience to teach them respect or fear of man. This makes them particularly dangerous to humans. As the South African naturalist Charles Astley Maberly noted, "Generally speaking, I think it is fear and distrust of the human scent, rather than dislike of human flesh, that prevents the majority of lions from attacking man deliberately" (1963, 155). Once recurring exposure and opportunity breed an element of familiarity with humans, this check on a predator's behavior is lifted and an attack becomes possible or likely. But even this predisposition does not make a habitual man-eater.

Transitions from Scavenging

Scavenging provides repeated occasions for the development of man-eating via opportunity. The clearest examples involve spotted hyenas. In many parts of Africa, hyenas provide a convenient and effortless means of garbage disposal. People find it much easier to simply throw their garbage into the brush outside town, and rely on the many animals that obtain food by scavenging. Hyenas are often the most numerous and important of these, and each night, animals patrol the outskirts of towns and even deserted village streets looking for a free meal. In Harar, Ethiopia, while researching his behavioral classic, *The Spotted Hyena,* biologist Hans Kruuk found that people not only tolerated hyenas but even encouraged them to feed on refuse. Only under the most dire circumstances do the hyenas prey on livestock. Frequent encounters breed reckless routines and relaxed vigilance on the part of people toward confrontations with one of Africa's most accomplished predators. Sometimes there are tragic consequences.

F. A. Balestra described an outbreak of man-eating in the Mulanje District of southern Malawi that may illustrate how this familiarity can prove fatal. In all, some twenty-seven persons died, all as a result of hyena attacks. The first reported victim was an adult man—the village idiot, who was attacked on the road between two villages. A week later and 8 miles away, an old woman was dragged out of her hut and fatally mauled before her neighbors could rescue her. A sleeping child was also attacked and eaten that year (1955), and in succeeding years, the hyenas claimed annual tolls of five, five, and six people. Each year, the killings began in September, when the climate is very hot and people sleep in unprotected locations, and ended in January, when people resume sleeping indoors. Interestingly, no reports of stock raiding accompanied these attacks (Balestra 1962).

Although the Mulanje episode was perhaps unparalleled in its long-term toll, such attacks are an everyday occurrence in rural Africa, as Kruuk's monograph chronicled. The *Tanzania Standard* for 22 January 1968 wrote, "Hyenas have bitten more than 60 people recently at Loliondo, northwest of Arusha, some in their houses, veterinary officials reported. The victims were mostly women and children. Hyenas were also reported to be terrorising an area just over the border in Kenya." And such attacks continue. Africa News Online recorded that in early June 2000, hyenas killed and ate at least four people around Sanaag, Somalia. The attacking hyenas snatched one victim, a four-year-old boy, from the arms of his grandmother. A local newspaper viewed these as the first recorded reports and attributed them to displaced animals that had come from eastern Ethiopia to escape the drought.

Teddy Roosevelt also remarked on man-eating hyenas: "Carrion-feeder though it is, in certain places [hyenas] will enter native huts and carry away children or even sleeping adults; and where famine or disease has worked havoc among a people, the hideous spotted beasts become bolder and prey on the survivors." Roosevelt described an outbreak of sleeping sickness in turn-of-the-century Uganda that had killed hundreds of thousands of natives: "In 1908 and throughout the early part of 1909 they grew constantly bolder, haunting these sleeping sickness camps, and each night entering them, bursting into the huts and carrying off and eating the dying people.... The men thus preyed on were sick to death, and for the most part helpless. But occasionally men in full vigor are attacked" (1909, 69–70).

Many African people forgo burial of the dead. Sometime this happens under the belief that the practice is harmful to the soil, but more often it occurs because the scale of mortality from disease, warfare, or famine is too massive to accommodate through normal burial or cremation. John Hunter (1952) noted the propensity for hyenas to act effectively as undertakers and thought this custom encouraged the animals to attack and kill humans. Some tigers may also learn to associate humans with food if they've had the opportunity to scavenge human corpses. The section of this chapter on funeral rituals revisits this topic.

Fewer examples of lions fit this profile, but this is only because lions are presumed to be predatory and their actions are typically interpreted in this light. Zoologist Brian Bertram, formerly a curator with the Zoological Society of London and longtime Serengeti researcher, described an incident that occurred at a village near Tanzania's Lake Manyara National Park. There, a lion frequented the town dump, feeding on scavengers and garbage. Over time, his prolonged exposure to human smells evidently caused him to lose all fear of humans. One night, this lion attacked a drunken, stumbling villager and killed and ate him. This lion ate two more people over the next few months before he was identified and destroyed. Even domestic dogs and pigs will prey on humans under such circumstances—a fifty-two-year-old South African man was recently killed by dogs and then eaten by pigs while returning home intoxicated.*

Old and Infirm Lions Attack Weaker Human Prey

Without question, old age and chronic infirmity are the most frequently cited causes of man-eating by lions and other big cats. Deprived of the means to attack or kill normal prey, predators frequently turn their attentions to slower, weaker, more defenseless victims—often people and domestic stock. Both represent nonnatural prey whose protection lies mainly in human traditions, customs, and retribution. The perpetrators are called "man-eaters" and "marauders" or "stock raiders."

Indeed, man-eating habits are so frequently linked with age and infirmity that it is impossible to determine where the idea first originated. Ninety years ago, Roosevelt and Heller observed, "As has long been known, man-eating lions are frequently very old individuals, males or females, which have lost many teeth, and are growing too feeble to catch game, whereas they find it easy to master man, who is the feeblest of all animals his size, and the one whose senses are dullest, and who has no natural weapons" (1914, 178). Almost surely, different people have independently reached this interpretation at different times about different species. Rarely is any justification given for the proposed cause-and-effect relationship.

A few examples will show the conviction that accompanies most of these pronouncements. The South African zoologist F. W. Fitzsimons stated, "In many parts of Africa where the villagers' goats and other animals begin to mysteriously disappear, they say: 'Ah, there is an old toothless lion about. We must turn out and kill him, or he will soon begin eating us'" (1919, 105). Similar reports include Armand Denis' observation that "the period when strength and speed are failing and teeth are worn is the danger period for man-eating" (1964, 76), Astley Maberly's contention that "old or disabled lions may take to pouncing on humans outside their huts at night, and so learn how easily they can be caught" (1963, 155), Maitland Edey's assertion that "it is the inability to hunt properly—often because of old

*The man's thigh and stomach had been consumed. http://www.igorilla.com/gorilla/animal/1999/dogs_kill_pigs_eat.html

age—that may turn a lion into a man-eater" (1968, 77), and John Hunter's determination that "most man-eaters are either old beasts that cannot hunt wild game or lions that have become injured in some fashion" (1952, 205).

All of these accounts were developed around African lions, but they are virtually identical to those proposed for marauding jaguars in American rainforests, man-eating tigers in Indian jungles, and leopards wherever they occur. The famous tiger hunter Jim Corbett wholeheartedly embraced this explanation, and the vast majority of his accounts identify broken teeth, festering sores, or other injuries that would prevent a cat from normal pursuit, capture, and killing of prey (1946, 1955).

Aging brings a seemingly inevitable weakening of life-support systems and diminution of physical, sensory, and perhaps even mental faculties. In carnivores, aging is manifested by exaggerated tooth wear. Older lions frequently have worn their incisors and some premolars down to mere nubs, exposing their pulp cavities to agents of decay and infection.* Some have also lost or broken teeth through injuries and accidents. Other evidence of a lifetime spent preying on large herbivores also becomes evident. Injuries that successfully healed—sharp stabs of an antelope's horn or sublethal kicks from a zebra or giraffe—begin to grow arthritic and develop "rheumatism." Eyes that once could follow the unsteady movements of a wildebeest calf on a moonless night are now clouded by cataracts. The territory that furnished food for years now lies in the possession of a younger, more powerful individual. And the gnawing hunger proves unbearable.

In South Africa's Kruger National Park, an old leopard recently killed a park ranger after being forced out of its traditional hunting area by younger challengers. The eleven-year-old cat was desperately hungry, observed the park director, David Mabunda. The old male was later found to have mange as well as numerous bite and claw wounds from other leopards. The ranger died almost immediately after the leopard leaped on him and grabbed him by the throat. He had been guarding a group of tourists near a bridge at nightfall (Arenstein 1998).

Animals in declining condition are continually subjected to higher risks. They may be forced to scavenge from the kills of other predators or groups, running a high risk of aggression for a scrap of food. Unable to catch and subdue fleet deer or antelope, they may turn to hunting porcupines. Although delicious and large enough to be rewarding, African and Asian

*A tooth is composed of a decay-resistant enamel coating surrounding a main body of dentine. This complex of living tissue is sustained by nerves and blood vessels that issue from the jaw, enter through the tooth's root, and pass into the tooth's interior pulp cavity.

porcupines can seriously injure lions and other big cats, as Fritz Eloff has described (1973). Whenever porcupines encounter a threat, they raise their quills in a threat display. This more than doubles their apparent size, and it also mobilizes a defensive armature far more imposing than those of their American counterparts. The quills of African porcupines may reach a foot in length. Escalating threats include foot stomping, rattling of the quills on the rump, and finally a wheeling backward charge, which drives the short, stout quills covering the lower spine into their would-be assailant. This defense is known to have severely injured or killed every kind of large predator in Africa, and lion encounters are common (Kingdon 1974; Nowak 1999). Festering porcupine quills have crippled both lions and tigers that subsequently became man-eaters, as documented by Jim Corbett (1946) and James Stevenson-Hamilton (1947).

It is ironic that humans themselves are often the cause of the injuries that subsequently create man-eaters and marauders. This association has long been appreciated—in the ninth century BC, Homer said of Diomedes, after being emboldened by Pallas Athene, that "he was three times as bold as he had been before, like a lion that a shepherd in charge of the woolly sheep on an outlying farm has wounded as he leapt into the yard, but failed to kill. He has only roused him to greater fury ...."* Although the weapons used to safeguard our villages and flocks have changed, the results have not. In India, according to Kenneth Anderson, "a tiger or panther is sometimes so incapacitated by a rifle or gun-shot wound as to be rendered incapable, thereafter, of stalking and killing the wild animals of the forest—or even cattle—that are its usual prey. By force of circumstance, therefore, it descends to killing man, the weakest and puniest of creatures, quite incapable of defending himself when unarmed" (1955, 8). Even without guns, human poachers can transform normal predators into man-eaters. Astley Maberly observed that most lion attacks on people in Kruger National Park have been traced to "the discovery of broken-off nooses of wire-snare deeply embedded in the necks of the poor brutes, causing unspeakable agony and gradual starvation" (1963, 155). Incapacitation of large predators greatly increases the risks that are inherent in human-carnivore coexistence. Victimized predators are unable to maintain their natural range against competitors or to forage on native prey. Chapter 9 discusses animal-human conflicts and their resolution in more detail.

Forensic Examinations: The "Smoking Gum"

Other explanations for man-eating (described in later sections of this chap-

*Homer. *The Iliad*. Book V, 95-96.

ter) tend to be strongly contextual and are very difficult to evaluate without a host of ancillary information. However, the age and infirmity hypothesis can be evaluated in specific cases simply based on the man-eater's remains. In the early 1990s, Field Museum preparator Tom Gnoske noted that one of the Tsavo lions had a broken canine and associated jaw malformation. Technical assistant John Phelps subsequently described the condition of this specimen in detail on a 1995 specimen invoice. These observations served as the basis for the statement of Kerbis Peterhans and colleagues that: "One of the man-eaters at the Field Museum had a broken lower right canine with an exposed root; asymmetrical growth of the skull in response to this abnormality suggests the beast had suffered from this condition a long time. Perhaps he was too disabled to hunt and consume the usual prey" (1998, 12).*

To refine these observations and to consider the consequences of this pathology, I enlisted the help of a specialist . Dr. Ellis J. (Skip) Neiburger is a practicing dentist in Waukegan, Illinois. He is the editor of the *Journal of the American Association of Forensic Dentists*, a past deputy coroner of Lake County, Illinois, a curator emeritus in anthropology at the Lake County Museum, and a prolific author. He had previously collaborated with William Turnbull, Field Museum's emeritus curator of fossil mammals, on studying a bear skeleton and Turnbull identified him as the collaborator I needed to extend existing analyses of the dentitions of Tsavo lions.

Neiburger and I reexamined the remains of the two historic Tsavo man-eaters and the recently acquired "Man-eater of Mfuwe" for possible traumas. The following account relies heavily on my collaborative studies with Neiburger, technical versions of which have appeared in *Nature Australia*, *General Dentistry*, *New York State Dental Journal*, and the *Journal of the American Association of Forensic Dentistry* (Neiburger & Patterson 2000a, 2000b, 2002, Patterson & Neiburger 2001). Here I would like to once again acknowledge my gratitude and appreciation for Skip's expertise and insights.

In all three cases, our forensic evaluations were necessarily limited to injuries to the skulls, jaws and skins. Only these parts of the cats were preserved. Therefore, it was impossible for us to determine whether post-cranial traumas, such as arthritis, broken limbs, or crushed vertebrae, may also have shaped their behavior, and all have been implicated as being contributing factors in other cases of man-eating. However, to our surprise, we

*Shortly after publication of the *Natural History* article, in early 1999, the initial four-way collaboration that Kusimba, Gnoske, Kerbis Peterhans, and I had established on Tsavo's natural and cultural history split into three teams, each independently pursuing research programs. I remained focused on the morphology, genetics, ecology, and behavior of Tsavo's lions.

found that all three cats had sustained important dental damage, and in at least one of them it was severe enough to render the infirmity hypothesis perfectly plausible.

Teeth are a vital part of the basic equipment of any mammalian predator. They permit these hunters to subdue and kill large, struggling prey and to process quickly and efficiently the large volumes of food that are needed to sustain their great energy demands. All three lions exhibit tooth and jaw damage that was probably sustained during normal hunting behavior. Hunting is a dangerous business, as UCLA paleoecologist Blaire Van Valkenburgh has shown. She quantified incidental breakage of teeth for various species of carnivores in several major mammal collections. Nearly one in four lions she examined had at least one broken tooth. In lions and in other species, canines were broken more frequently than other teeth, followed by the powerful shearing cheek-teeth known as carnassials (1988).*

The risk of breakage to teeth follows directly from their function. The great cats use their canines ("eye-teeth") to stab, hold, and kill. In the case of large prey, lions typically seize the throat and suffocate the prey with a throttling grip, but with smaller, more manageable prey, they will bite through the nape of the neck with a killing bite to the brain stem or else sever the spinal cord. To serve effectively in either case, canines must be long and thin. As Van Valkenburgh has determined, this design leaves them vulnerable to snapping from tensile forces as prey continue to struggle (1988). On the other hand, carnassials of lions are designed for quickly shearing the large volumes of meat that lions routinely ingest at kill sites; occasionally, they are also used for crushing,† to extract marrow from cracked bones. The prevalence of broken teeth in Van Valkenburgh's survey and its costs in fitness terms are proof of the powerful selection pressures that shape the evolution of mammalian teeth. They also illustrate the perils of preying on large animals. Her studies of Pleistocene faunas show that tooth breakage is an age-old problem for large carnivores (1993).

Because tooth breakage is common among carnivores but man-eating is rare, the existence of a broken tooth is inadequate evidence for a claim of

*This is actually a functional complex, consisting of the upper fourth premolar and the first lower molar, typical of all members of the order Carnivora. It assumes different shapes and functions in different members of this diversified group.

†Van Valkenburgh also showed that lions exhibit heavier wear and higher frequencies of breakage than cheetahs, presumably because the latter seldom gnaw on bones. Also, female lions exhibited heavier wear and higher incidence of breakage than males, owing to the greater numbers of older females in the sample (Van Valkenburgh 1988).

crippling infirmity. Dr. Neiburger and I believe that it is the *disease and pain that are sometimes associated with tooth breakage,* not breakage itself, that may be responsible for altered behavioral patterns and more probable conflicts with humans. Since our initial study, we have begun a broader survey to more fully document the disease that accompanies tooth breakage.

The First Tsavo Man-Eater. Back in Tsavo, when Col. Patterson approached and inspected the body of the "first man-eater," he immediately noted a dental injury. He stated, "On examining his head, I found that a .303 bullet had smashed out one of his tusks, for the track of the bullet was left in the tooth stump. I must have given him a bad toothache the night he attacked Brock and myself in the freight car" (Patterson, 1925, 34). However, the cleaned skull shows extensive wear on the stump of the broken right lower canine, and this degree of wear takes years to develop (see Figures 6 and 7). The bullet course that Patterson spotted in the axis of the tooth is in reality the pulp cavity, a hollow chamber in the center of the tooth that carries nerves and blood vessels to nourish the living tooth. In the first Tsavo man-eater, this chamber was enlarged and riddled with decay from trapped food particles that had decomposed there. As Patterson noted, its lumen was now about the size of a bullet a third of an inch across.

Adjacent to the broken canine, three lower incisors (unicuspids, or nipping teeth) are missing. The two inner incisors are missing and have vacant, smooth-walled sockets; on these grounds, they might have fallen out while the skull was being cleaned. However, the third, a comparatively massive and powerful tooth in most carnivores, is missing and its socket mostly filled with spongy bone. The tooth was lost long before death, for the bone was laid down during healing, the lion's repair mechanism to prevent further infection and decay in this exposed, gaping cavity.

The teeth of the upper jaw provide other clues that this lion broke its lower canine long before his death. The three upper right incisors are all "super-erupted," meaning that they grew beyond their normal lengths, something that happens only in the absence of contact with the opposing teeth. Again, this additional growth implies an extended period, at least a year or more, after the trauma and prior to death.

A precise articulation of the powerful upper and lower canines acts to control jaw conformation. The lower canine ordinarily acts as a spacer, maintaining the position and orientation of the upper one, which is tucked tightly behind it in the toothrow. In this case, however, the upper right canine had rotated forward and inward, so that its tip came to rest on and actually abraded the stump of the broken lower tooth, rather than extending some inches beyond it as is normally the case. The abnormal, dysfunc-

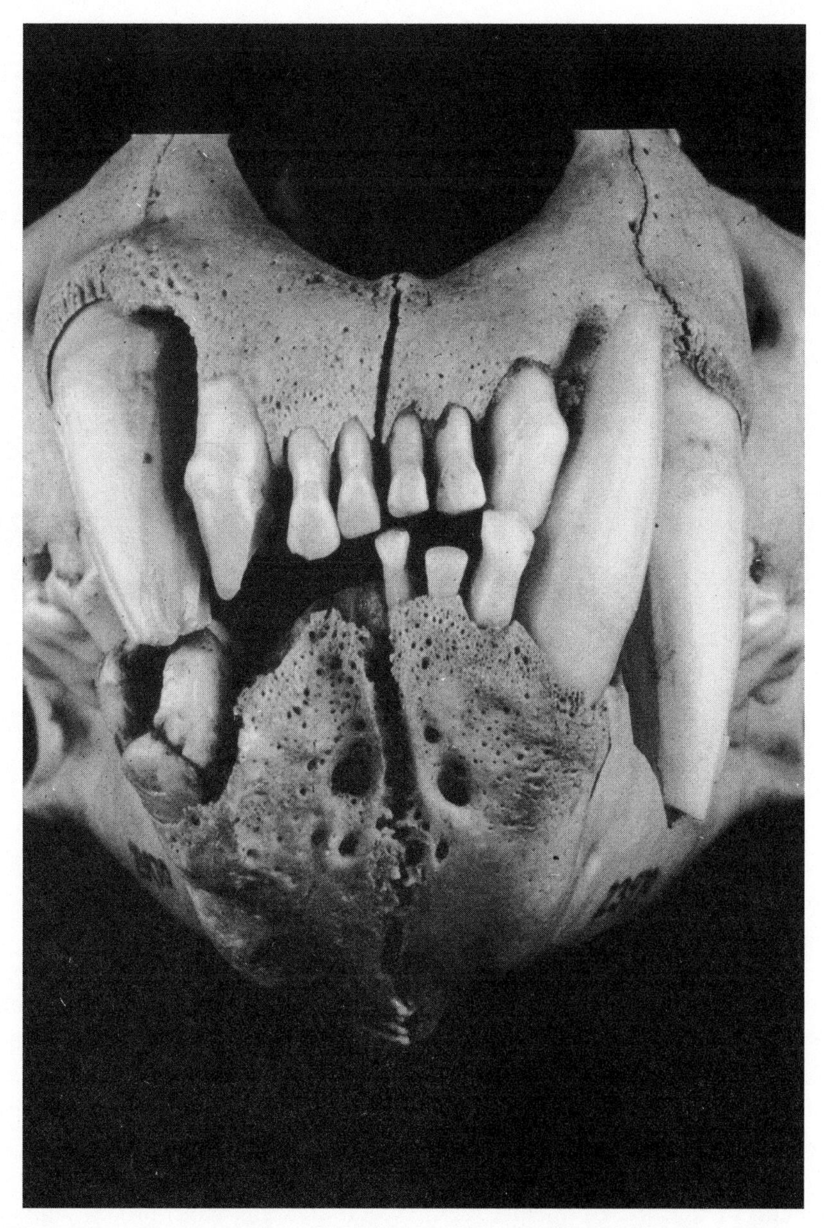

Figure 6. Articulated jaws of the "first man-eater"of Tsavo. Broken lower right canine is visible at left, as is super-eruption of upper right incisors and rotation and malocclusion of upper right canine. Cranial asymmetry of this individual developed after its initial injury and prior to its death. Photo by J.S. Weinstein and B. D. Patterson. (Field Museum neg. # Z94320_11c, © The Field Museum.)

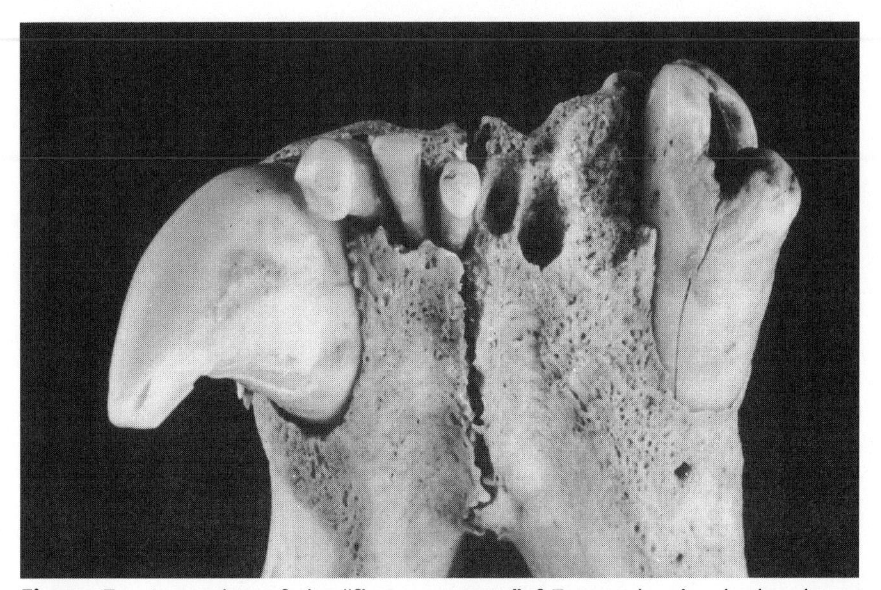

Figure 7. Lower jaw of the "first man-eater"of Tsavo, showing broken lower right canine with exposed pulp cavity. Socket of neighboring third incisor is filled with spongy bone. The root-tip abscess that Neiburger and Patterson discovered with x-ray images is located inside the body of the right jaw, between the canine and the first premolar. Photo by J.S. Weinstein and B. D. Patterson. (Field Museum neg. # Z94321_11c, © The Field Museum.)

tional occlusion of the canines accounts for the extensively worn tip of the broken canine. Again, this rotation and malocclusion of the canines must have developed over an extended time period.

Most significantly, radiographs taken in Neiburger's Waukegan offices show that the broken lower canine had a large root-tip abscess. Such abscesses are commonly associated with significant pulp exposures, as the blood vessels and nerves in pulp are full of nutrients for bacteria and become fertile grounds for infection. The abscess of the first man-eater suggests that this condition was very painful, as well as awkward and dysfunctional.

In the case of Tsavo's first man-eater, poor occlusion limited the effectiveness of the lion's principal stabbing and grabbing teeth on one side of the mouth. In addition, the cat would have experienced excruciating pain with any pressure on the tip of the broken canine. Under these circumstances, it would have been impossible for him to apply the typical "killing bite" maneuver. Although buffalo may be favored prey for many lions today, they would have been far too dangerous for this lion to tackle—and if you can't dispatch a buffalo, you'd better not grab one, because it surely could dispatch you! The first man-eater clearly had to adopt new rules for hunting to compensate for his losses, and his large size and good condition showed he had been successful at this. Dental problems alone would have rendered the territory of Tsavo's first man-eater a very bad campground indeed for either railroad workers or slave caravans.

The Second Tsavo Man-Eater. The teeth of the second lion that Patterson shot were in far better condition, although the skull itself was in much worse shape. In his 1925 account, Patterson describes firing at least ten shots at this cat, hitting him with at least six shots. One shot broke the massive zygomatic arch (cheekbone) on the left side and buried itself in the inner wall of the eye socket. Other shots shattered the lower jaw, so that the left side is missing shortly behind the toothrow and the right side is also damaged. But all that damage happened at the time of death.

The only dental malady we noted in examining this individual was a fracture on the surface of the upper left carnassial. In two places, the pulp cavity had been exposed, but neither showed indications of a root-tip abscess or other pathology. Because the broken surface was mostly unworn and uninfected, we believe that this breakage was quite recent, perhaps within a few months of the lion's death. While the pulp exposures could have caused its owner some discomfort while chewing, lions routinely tolerate pulp exposures of equivalent size on their incisors and canines as they become worn with age.

Did the "slab fracture" on the carnassial of Tsavo's "second man-eater" drive this cat to man-eating? This seems unlikely, given the frequency of damage to this tooth among free-living lions. Van Valkenburgh's survey suggests that most animals with broken teeth should behave normally; almost a quarter of the museum samples she examined had fractures like that of the second man-eater, so "normal behavior" should often be associated with dental damages. As noted earlier, it is the atypical, pathological disease sometimes associated with broken teeth—not broken teeth themselves—that may precipitate man-eating behavior.

Lions are intensely social animals (as discussed in Chapter 5, "Lion Biology"), and both males and females form lifelong associations with both kin and unrelated individuals. Patterson's many accounts of the attacks show that the two lions hunted and foraged cooperatively. In fact, in November 1898, near the end of the "Reign of Terror," Patterson stated, "Hitherto, as a rule, only one of the man-eaters had made the attack and had done the foraging, while the other waited outside in the bush; but now they began to change their tactics, entering the *bomas* together and each seizing a victim" (Patterson 1907, 69). The first man-eater was killed about two weeks afterwards on 10 December 1898, so that the second lion had little time to learn and perfect its skills. In the three weeks it took Col. Patterson to locate and dispatch the second lion (which fell 29 December 1898), it prowled settled portions of the camp and killed and ate camp goats but failed to claim any human victims. Because the long string of killings ceased with the death of the "first man-eater," I presume that it was the second that had customarily waited patiently in the bush.

My own belief is that the second lion was guilty by association—his social bonds swept him into his nefarious career. Unfamiliarity with the camps and countermeasures of men may have led to his comparatively speedy demise; alternatively, the shotgun wound he received from Patterson over the goats earlier used as bait may have partially incapacitated him. But this interpretation raises the ancillary question, "Where was the second man-eater as the attacks took place?"

The Man-eater of Mfuwe. Neiburger and I also had access to a third specimen of a man-eater, the one Hosek killed in Zambia and donated to the Field Museum. The teeth of this lion were normal for an animal of about five years of age, mostly unworn and without cracks or pulp exposures. However, the right side of the lower jaw had a system of major injuries along its lower margin. Radiographs confirm that tooth disease was not responsible for these lesions—they were probably produced by blunt trauma, such as the kick of a large hoofed mammal. In support of this interpretation, we also found minor damage—resulting from inflammation of the membrane over the bone—in the jaw socket of the cranium on the opposite side of the skull. A kick to the right side of the face would have forced the jaw to the left, transmitting the force to the jaw socket on the left side. The three lesions on the right side of the jaw have rounded margins and communicate with each other and with the mandibular canal; this conduit for nerves and blood vessels that supply the front of the jaw has been partially or completely obstructed by the healing process.

Ordinarily, and on the left side of the Mfuwe man-eater's jaw, the canal passes from an opening (foramen) on the inner side of the jaw near the joint to exit from another one just behind the canines on the bone's outer surface. A laser probe inserted into the undamaged canal passes easily from one foramen to the other. Inserted into either foramen on the damaged side, the same probe is obstructed by the overgrowth of bone that was laid down in response to the trauma.

Bony repair of the initial trauma appears well healed, but this condition likely had lingering functional consequences. There is evidence that the open lesions were probably pustulous and drained either into the mouth or externally onto the jaw. After Neiburger and I detected this anatomical condition in 2000, I asked Wayne Hosek about the cat's appearance, and he confirmed that an assistant had remarked in the field that the Mfuwe man-eater "looks a little green around the gills,"* presumably from pus associated with this injury. Such infections may commonly be responsible for the "stench" of a man-eater's breath (see "Kima Killer" in Chapter 1).

There can be little doubt that the Mfuwe lion had sustained severe blunt trauma to the jaw. His recovery from this injury required an extended period, a period substantially longer than the two months he is thought to have preyed on humans. Healing almost certainly left him with some impairment, as both the nerve and the blood supply to the lower jaw were interrupted altogether or rerouted from the obstructed mandibular canal. The Mfuwe man-eater probably sustained his crippling injury well before he began killing his documented victims, and those depredations started well after he was on the road to recovery.

However likely that dental traumas played a role in predisposing both Tsavo and Mfuwe lions to their man-eating habits, it does not follow that all man-eaters have bad teeth. In fact, my colleagues and I recently compared dental traumas and disease in various museum samples of lions with those in 22 "problem lions" shot outside the Tsavo parks for attacking livestock or people (Patterson, Neiburger & Kasiki 2003). If the "infirmity hypothesis" explained all instances of man-eaters and marauders, we could expect a higher incidence of traumas and lesions in the Tsavo group. Instead, statistical analyses showed that the problem lions had *better* teeth! It turned out that most problem lions were younger on average, and had sustained less dental injury that most lions shot by sports hunters (the source of most museum samples). Born in the park and dispersing outside it to find unoccupied territories and uncontested food, these lions found themselves in human settlements and rangelands, where their predatory habits soon led to conflicts with people. This example offers a natural segue to the effects of prey availability.

Prey Availability and Attacks on Humans

At the end of 1898, ... the surroundings of Tsavo looked very different from what they are now.... Game was remarkably scarce, Patterson

*Conversation with BDP, 15 August 2000, over lunch at the Chicago Park District Headquarters, McFettridge Drive, Chicago.

> *recording water-buck along the rivers, dikdik, lesser kudu, and rhino in the bush. The lions must have greeted the sudden appearance of large crowds of railway workers with considerable enthusiasm. Here, at long last, was a rich supply of food, of easily obtainable food at that.*
> (C. Guggisberg, *Wild Cats of the World* 1975, 167)

Lions are great opportunists. As with any other animal, a lion's evolutionary goal is to leave as many descendant lions as possible. However, the lion is an apex predator, and this has some interesting consequences. First, because it occupies the peak of its food pyramid, lions can't afford to be too choosy or to specialize in any one type of prey. Such tight linkages are frequently seen at lower levels in the food pyramid, as between herbivorous insects and their host plants. Lions must rely on whatever acceptable prey species are abundant enough to support a viable population of lions, and this is apt to change from place to place and from time to time. If no single preferred species is abundant enough, as is usually the case, then several kinds of prey may be chosen interchangeably. Second, lions are only rarely excluded by other species from their foods. More frequently, it is lions that interfere with prey selection by other species, stealing carcasses from competing cheetahs, hyenas, and wild dogs. This means that lions are free to exercise their dietary choices without much active interference by other species.

Where prey is abundant, lions can specialize in one or a few preferred prey species, but where food is scarce, they search far and wide and feed on a greater diversity of prey. As Mervyn Cowie put it, "In this way it can happen that a relatively young and healthy animal can turn into a man-eater just because it has chanced upon its first victim by accident, after which he finds this method of hunting very much easier although ultimately more dangerous" (1966, 61). In this case, a lion actively chooses to hunt and kill humans, and the behavior is quickly entrained because humans are abundant and easy to capture.

Prey scarcity demands that lions be opportunistic in their diets. This simple fact suggests two testable hypotheses about man-eating. The first has to do with the role of sociality. Because a lioness lives her entire life in the company of kin, potentially helpful daughters, nieces, and cousins surround her into her dotage. Kinship might offer lions a form of "social security"—sharing food at communal kills (or at least scavenging from their leftovers) and being tolerated on their territories. This bequest to old-timers would be correspondingly larger for females than for males, so that one might predict that a higher proportion of males might become man-eaters. To my knowledge, this question has never been investigated over a large enough sample to be statistically valid. Unfortunately, the anecdotal literature has not been assembled in a comprehensive and regular fashion that would permit reliable analyses.

Similarly, no systematic comparisons of man-eating have been reported between lions that are resident on territories and nomadic lions. Simply being able to maintain a residence is an important statement of health and vitality in lions, and tenure must be a useful measure of habitat and resource suitability. One would predict that resident lions should less often become man-eaters or marauders for two reasons: first, they should reside in relatively high-quality habitat better stocked with native game; and second, such areas should be better insulated from human effects. Again, I am unaware of prior attempts to test this idea, but anecdotes seem to support it. In October 1999, the *Star*, a South African newspaper, described steps being taken to safeguard citizens from a small spate of attacks—five in a single year—that took place near Marloth Park on the border of Kruger National Park. Two lions, both nomadic males, were known to have consumed a person in town and were slated for relocation. Managers were using radio collars to determine whether other lions in the area had altered their space-use behaviors so as to pose a danger to humans. Wild animals are allowed to roam Marloth,* but because it lies outside the national park, game is obviously scarcer and hunting by lions more difficult there.

Any circumstances or events that alter the abundance or composition of prey species can potentially influence a lion's diet and affect its likelihood for becoming a man-eater. Both natural and artificial causes can trigger such changes. Below, for purposes of organization, I offer separate discussions of lion habits where prey is ordinarily scarce and their habits in the face of sudden scarcity caused by epidemics, droughts, or over-hunting. Here, I emphasize that both have a common basis, and often a common consequence, in the "optimal foraging" decisions by lions.

Regions Where Game Is Naturally Scarce

> *Most, though not all, of these districts [where man-eaters prevail] consist of dry thorn-bush which never holds much in the way of game, though there may be plenty of game not so very far away in more suitable country.* (J. Taylor, *Maneaters and Marauders* 1959, 16)

As highly adaptable animals, lions respond to changes in the abundance or

*http://wildnetafrica.co.za/bushcraft/dailynews/1999archive_10/archive_19991007_ lionstobecollared.html

availability of prey by "prey switching," substituting one acceptable species of prey for another. On a geographic scale, the diets of lions vary enormously from place to place, reflecting this adaptability. Even neighboring prides may eat very different prey species owing to local differences in abundance, availability, or experience. Lions respond in a similar manner over time to changes in prey abundance or availability at any one place. University of Minnesota researchers David Scheel and Craig Packer studied prey selection by individual prides of Serengeti lions over the course of the migration, finding that levels of predation on each ungulate species vary between seasons, across habitats, and from year to year. The annual migration of wildebeest, zebra, and gazelle is responsible for much of this variation. Ease of procurement, nutritional requirements, and taste preference are the factors that determine what is optimal for any pride at any time. (This topic is considered in much greater detail and with greater substantiation in Chapter 5, "Lion Biology.")

In general, predators take a range of prey items that shrinks as the concentration and local abundance of suitable prey increases. That is, where prey species are highly abundant and clumped, lions can afford to specialize; where they are scarce and scattered, lions and other cats must be opportunistic generalists. The same pressures that force a lion living in unproductive habitat to seize hares and dik-diks may cause it to attack livestock and herdsmen as well. For tigers, prey is routinely scattered across the dense, forested areas they inhabit, and this might explain why tigers will attack virtually any vulnerable animal that they encounter. In such circumstances, "selectivity" disappears—the predator takes prey of all ages and physical conditions in approximately the same proportions that it encounters them.

This rationale makes good sense and is now part of conventional wisdom. For example, during their survey of crocodiles, Alistair Graham and Peter Beard had a very simple, unprotected camp at Moite, on the eastern shore of Lake Turkana. One morning they found lion tracks leading to a few feet from where they had slept, which they found completely unnerving: "The lion was apparently only inquisitive, but I did not trust it, for in northern Kenya maneaters are common. On the comfortable plains of Masailand, where food animals abound, the lions are good-natured and leisurely, with time to grow luxuriant manes of long black hair. But in this arid, hard desertland they are fierce and unmerciful, for they must exploit every opportunity" (1973, 65).

The trigger for man-eating by this mechanism is a pervasive, recurrent habitat feature, not one that depends on a cat's individual experience; accordingly, man-eating becomes associated with a particular region. While acknowledging that man-eating is in general rare, Astley Maberly allowed that it was far commoner in some areas than in others. John Hunter too emphasized this geographic link: "There seems to be a tendency among lions in certain districts to become man-eaters—a peculiar hereditary taint that cannot be explained.... It occurs especially in the Tsavo district which has been famous for its man-eaters ever since 1890 [*sic*]. Today, most of the lions have been shot out of this section, yet there still are occasionally reports of an odd man-eater" (1952, 205–206).

John Taylor probably developed this habit-habitat connection more fully than any other writer, as indicated by this section's epigraph. Although he wrote mainly of Mozambique ("Portuguese East Africa"), he provided a precise description of the principal vegetative formations around Tsavo—dense thorn-scrub that in places is so thick that it is passable on foot only along rhino trails. These areas tend to be sparsely settled by people and thinly stocked with both game and livestock—much of it has little grass cover, and malaria, tsetse flies and sleeping sickness are endemic. The limits of this habitat zone stretch from the Zambezi River in Mozambique to north of Lamu along the Kenya-Somalia border. Although he acknowledged a few gaps in the distribution of the thorn-bush, Taylor regarded the formation as largely continuous. Within this zone, he identified regions chronically plagued by man-eating: the notorious coastal belt, the lower Zambezi valley (Mozambique), others in southern Malawi and near the lake, the entire Lindi province and Usorii district in Tanganyika, and the "*nyika*," the barren, treeless wilderness of eastern Kenya and adjacent countries. "Throughout this entire coastal belt you will find maneaters. Some stretches are worse than others; but it can safely be stated without fear of contradiction that no stretch is ever entirely free from the scourge for more than a short period" (1959, 105).

The foregoing also describes the habitat structure and prey distribution of the infamous "Man-eaters of Njombe," although that district lies far from the Indian Ocean. Between 1941 and 1946, 15 lions killed at least 246 people in this one district; because they terrorized three districts during this period, a conservative estimate of deaths caused by the lions could be 1000 or 1500. George Rushby, their nemesis and chronicler (like Patterson in the case of Tsavo, perhaps not an entirely impartial judge) claimed, "The renowned man-eaters of Tsavo were very small fry compared to what these proved to be" (1965, 185). Certainly no other set of lions have been charged with so many killings. Like Tsavo, the territory of the Njombe man-eaters was dominated by bush and was "not game country."

Taylor so fully subscribed to the habitat-based interpretation of man-eating by African lions that he felt it necessary to flatly reject Corbett's "infirmity theory." He noted that his quarries included many examples of lions in good to excellent condition, the prime of life, and without injuries of any sort. Moreover, prey shortage—either on a regional basis or episodic—appears to elicit man-eating in tigers. In its fact sheets, Sea World/Busch Gardens states, "Tigers living in habitats where native prey is scarce may be forced to hunt humans for food. Likewise, a mother tiger with cubs may hunt humans to provide enough food for her young."*

Whereas an earlier cause of man-eating was identified as "opportunity," we might label this related one "exigency." Both depend on the adaptability and opportunism of cats. However, under the former cause, lions accidentally discover man-eating in the course of searching for more regular meals. In the latter, a "regular" meal is purposefully defined loosely: any prey large enough to be worth the effort and small enough to be subdued. For lions in habitats sparsely stocked with ungulates, a human might always qualify, depending on the time of day, risk of interference or retribution, and the like.

Some of the most convincing evidence that man-eating behavior may arise by this means was provided by an unwitting experimental demonstration! In the 1940s and 1950s, "game control" was used in certain districts to increase the suitability or possibility of settlement and commercial development in the face of tsetse flies. Wherever these flies (and the sleeping sickness they spread) were common, they effectively precluded such development over vast areas of east Africa. Thousands of wild hoofed animals were simply gunned down to eliminate them as hosts for the flies or as reservoirs for sleeping sickness. Writing of this episode in *The Tree Where Man Was Born*, Peter Matthiessen stated, "Lion attacks are now quite rare, but in the days of widespread game slaughter for tsetse control, a number of lions in these devastated regions turned on man in desperation, and bags of fifty, sixty, and in one case ninety human beings were recorded" (1972, 171).

Earlier in this chapter, we noted that human persecution of carnivores can actually increase the likelihood of man-eating, by crippling otherwise capable predators and making them infirm. In a similar manner, human activities that decrease the carrying capacity of land for game elevate the chances of man-eating via exigency. Habitat conversion to range-

*http://www.seaworld.org/infobooks/Tiger/diettiger.html

lands or farms serves to replace natural prey with livestock. Similarly, encroachment of agricultural lands into savannas heightens the spatial overlap of human and carnivore. Both shorten the chain of behavioral links necessary to establish and entrain man-eating. Unlike the case with a single maimed or wounded predator, *all* predators hunting this land will face these pressures for competition with humans, considerably elevating the odds of conflict. Charles Guggisberg concluded that "where game has been completely eradicated, lions come in close contact with man through killing domestic stock, and, although conservative in their predatory habits, some of them may eventually switch from cattle to herdsmen and to humans in general. A lioness that has made the killing of humans her specialty will naturally teach her cubs how to hunt people, and there may eventually be a real outbreak of man-eating. Let us repeat, however, that such incidents are most likely to occur in regions where man has shot out the game animals which used to form the lions' traditional prey" (1975, 166–167). Lamentably, such places are ever larger and more numerous in Africa.

Prey Scarcity Caused by Rinderpest Epidemic

> *Lions readily become scavengers, and with the plains littered by the carcasses of cattle, these big cats increased greatly in numbers. Weakling cubs that would soon have died under normal circumstances grew to maturity and thus in a surprisingly short space of time the Masai country was overrun with lions. When the epidemic had run its course and there were no more dead cows lying about, the lions turned on the live cattle.* (J. A. Hunter 1952, Hunter, 72, describing a rinderpest epidemic of the 1920s)

In the late nineteenth century, Africa was rocked by cattle plagues that decimated native herds, wrecked local economies, destabilized governments, and led to widespread die-offs of many wildlife species. The resulting shortage of prey may have caused lions and other carnivores to seek alternative foods to include in their diets. Many have suggested that the lions in Tsavo were driven to man-eating by prey depletion as a result of a rinderpest epidemic there. Rinderpest is an epidemic that infects both domestic livestock and hoofed wild mammals. By reducing the number of alternative prey animals, rinderpest may have elevated the potential importance and likelihood of man-eating. Hunter (above) identified the likely chain of causation: from dead cows to live cows to herdsmen to humans generally. Several earlier discussions document the intimate association of marauders and man-eaters. This "chaining" of behavior typifies not only lions in African savannas but also jaguars in Central American jungles, which seldom turn into man-eaters. Nevertheless, Wildlife Conservation Society zoologist Alan Rabinowitz avers, "It is believed that many man-eaters become cattle killers first" (1986, 202).

Rinderpest is an acute and highly infectious viral disease of cattle. Its causative agent is a morbillivirus, one of a notorious group of viruses that includes those that cause parainfluenza, mumps, measles, distemper in dogs, and Newcastle's disease in birds. Rinderpest is transmitted through the secretions and excretions of infected animals and enters through the respiratory tract of susceptible ones. Originally native to Asia, where it may induce only mild infections in susceptible animals, it is a scourge of many kinds of cattle and their relatives. The disease is so virulent in cattle that it is usually introduced to them by infected sheep and goats, which can harbor the disease a much longer time, spreading it to more susceptible species.

The clinical symptoms of the disease are heart-wrenching, even for a city dweller not overly fond of cattle. Animals with the disease exhibit congestion; watery discharges from the eyes and nose; loss of milk production; areas of dead tissue on the mucous membranes of the mouth, nostrils, and urogenital tract; and respiratory problems. A high fever often develops three to nine days after infection, with inflammation of the mucous membranes of the mouth, nostrils, and vagina and vulva. Tearing and drooling develop, with suppurating or bloody discharges, and lesions induce sloughing of both the nasal and intestinal linings. This in turn initiates further blood loss as well as diarrhea and dysentery. Collapse and death rapidly follow from massive fluid loss, dehydration, and subnormal body temperatures.*

Humans brought rinderpest to Africa. The first epidemic began in 1887 at Massawa, in what is now Eritrea, when Indian cattle infected with the disease were shipped there. The disease quickly spread through native herds of cattle and to wildlife, because much of Africa's big game, including all its antelope and buffalo, belong to the cow family, Bovidae. The ensuing epidemic claimed 95 percent of Ethiopia's cattle, causing four years of famine.

In an unpublished thesis at Northwestern University, Ben Polak developed a very detailed reconstruction of rinderpest's course in Kenya (1986). The disease reached the Maasai *kraals* near Naivasha in March 1891 but

*http://panis.spc.int/RefStuff/Manual/BOVINE/RINDERPEST.HTML, http://disaster.cprost.sfu.ca/epix/topics/animal/rhindpst.htm

did not reach southeastern Kenya until later that year, near what is now Kitui. Traveling south from Lake Victoria and the Kilimanjaro district, rinderpest passed through modern Tanzania, heading south but not west, so that Malawi was spared for a time. The epidemic raged south, crossing the Zambezi River to reach Bulawayo, in Zimbabwe, by October 1895. By early 1896, large numbers of both cattle and wildlife had died on both banks of the Zambezi, and most of the cattle around Salisbury (Harare) were dead within two weeks. Khama, a paramount chief of the Bamangwato tribe in what is now Botswana, may have lost more than eight hundred thousand head, according to Blayney Percival. It would be difficult to exaggerate the social and economic impact of such losses.* The disease overwhelmed Zimbabwe by March 1896 and reached the eastern Transvaal later that same year. It traveled over a thousand miles across South Africa in a single year, infecting lands from the Zambezi to Cape Colony. Its spread in southern Africa was speeded by the ubiquitous use of ox wagons for transport.

What effects did this rinderpest epidemic have on wildlife? Again, Polak's review is more complete than any single published source. On a railroad survey expedition across Kenya in 1892, Polak cites Pringle as writing, "Throughout the expedition no member of the survey party was fortunate enough to see a single buffalo, and only on two occasions were tracks of the animal to be met with...." Only two years earlier, these same areas had teemed with thousands of buffalo. Kudu and eland also succumbed to the disease in great numbers, so that all travelers saw of them were skulls and bones. Populations of these species recovered slowly during the 1890s and did not again become abundant until the first or second decade of the twentieth century.† The disease also affected gazelles, hartebeest, and zebras, although much less virulently.

From this abbreviated case history, it is apparent that rinderpest had run the course of Africa in seven or eight years. The rate and extent of infections varied substantially, depending foremost on the population density of cattle, other livestock, and wildlife and on their movement or migratory habits. Natural disasters like droughts also influenced its course of development. In any event, the early outbreak of rinderpest in Kenya (1890–1891)

*As one example, rinderpest, together with East Coast fever and adverse climatic conditions, probably speeded the movement of Zulus to the Witwatersrand during the first decade of the twentieth century, which had enormous impact on South African history (Ramdhani 1989).

†When Carl Akeley undertook a collecting expedition for the Field Museum to Kenya in 1905, elands were still protected from hunting by a $25 fine. Akeley was eventually given permission to secure buffalo specimens for the museum, but only in the Tana River district and not until July of that year, so carefully were surviving stocks protected (Bodry-Sanders 1991).

meant that even secondary and tertiary ripples had subsided by the time the railroad was passing through Tsavo. Almost certainly, the outbreak that originated in Eritrea in 1887 had very little to do with the Tsavo man-eaters in 1898. But it appears there was a secondary outbreak that might well have affected events in Tsavo.

The 1898 Epidemic. Late 1897 and early 1898 was a challenging time for the British East Africa Protectorate. Summer rains had failed for two consecutive years, causing water shortages and nearly universal crop failure. Trade and labor from India, including relief supplies of grain and other emergency goods, were interrupted when a quarantine was imposed to protect East Africa from the subcontinent's "Bombay plague" (see Chapter 1). In addition, pleuropneumonia, a cattle disease resembling rinderpest but generally causing lower mortality, had erupted between Machakos and Kitui. This confluence of major factors, several of which conceivably might affect man-eating, makes it difficult to ascribe causation to any one of them.

The protectorate's commissioner, Arthur Hardinge, stated in a report to Great Britain's House of Commons that in March 1898 some cattle infected with rinderpest had been shipped by rail to armed forces in Uganda (Hardinge 1899). The resulting epidemic quickly raced through the whole of Ulu, killing thousands of head near Machakos (SE of Nairobi). Despite a government decree to destroy the carcasses of dead cattle to prevent spread of infection, the Wakamba actually butchered carcasses for food, carrying the disease to unaffected villages. From a colonial standpoint, the effects could even be gauged statistically—revenue from livestock fell to one-quarter that of the preceding year, and the epidemic equaled in general intensity the outbreak of 1891–1892 (Hardinge 1899). J. A. Haslam, a second veterinary officer, arrived in May 1898 in response to the renewed outbreak—he was murdered later that year by Kikuyu herders who misunderstood and distrusted his incessant inoculations and postmortems of cattle. The epidemic eventually burned out in January 1899, having killed 40–50 percent of the cattle along the section of rail from Kitui to Machakos. Curiously, this epidemic coincided almost exactly with the man-eating events in Tsavo—the infected cattle arrived the very month that Col. Patterson began work in Tsavo, and the rinderpest epidemic abated a month after the man-eaters were shot.

The 1898 epidemic would have surely produced a surfeit of cattle carcasses on which the lions could feed, especially once rinderpest became established along the railroad as it blazed up-line toward Machakos (March through June). It is possible that this glut of easy meat offered Col. Patterson and company respite from the man-eaters from May through

September 1998. Lions might have exploited cattle carcasses nearby rather than feasting on undocumented rail workers "up-rail." However, once carcasses came into short supply, the "prey depletion" hypothesis may have explain the initiation of the final "Reign of Terror" that lasted until December of 1898.

Acute Effects of Drought: Human Corpses Become Food Items

> *Rain has been so long delayed that twelve months have elapsed since it has fallen in any quantity, and at the date at which I am writing nearly all the districts on the coast are in a state of famine having exhausted their locally grown grain, and depending for subsistence on that imported from India.* (A. Hardinge 1899, 6).

Although drought affected large areas of eastern Kenya in 1897 and 1898, its worst effects were felt by the Wakamba, living in an region that reached from the northwestern region of Tsavo nearly to Nairobi. Hardinge noted that Kitui was hit so hard that men were forced to emigrate to Kikumbuliu, 50 miles away, to seek employment on the railway, abandoning women, children, and the elderly to face the brunt of local conditions. Widespread crop failures forced residents to rely on emergency supplies from India. When the plague quarantine interrupted those shipments, famine began in earnest. A German, Dr. Kalb, traveling from the coast to Mount Kenya at this time, counted over one hundred corpses of natives who had succumbed along the roadside (Hardinge, 1899).

The severe impact of the drought on the Wakamba follows from geography and their economy. Historically, the Wakamba were a hunter-gatherer group displaced from the south, apparently owing to increasing expansion and militarism of the Maasai (Kasperson et al., 1995).* Once in southeastern Kenya, they settled in arid lands that were marginal and often inadequate for crop cultivation. Nevertheless, the tribe secured a dependable living from a diversified economy of hunting, herding, and farming. During the 1892 epidemic, the Wakamba were able to rely on agricultural production and hunting to compensate for rinderpest's toll on their herds. But the drought in 1898 eliminated this cushion, with severe ramifications for their economy.

Those ramifications were in immediate evidence at Tsavo Camp. A party of starving Wakamba happened to pass the camp one morning as

*http://www.unu.edu/unupress/unupbooks/uu14re/uu14re0k.htm

Col. Patterson sat down to skin a leopard shot for marauding the camp's goats and sheep. The desperate Wakamba offered to perform that job for him in exchange for the meat. In a very few minutes, they handed him an expertly skinned hide, then commenced to make a ravenous meal of the leopard, not even bothering to cook it. The combined effects of drought, which initiated the hard times, and of the second rinderpest outbreak, which removed the Wakamba's "insurance policy" against loss of agricultural production, were catastrophic. Coupled with suspended shipments of relief grain from India and the introduction of smallpox in 1899, this resulted in a great famine that lasted until 1901. Historians writing at the turn of the last century estimated that the mortality rate in the Kamba districts approached 50 percent. This and Kalb's reports of hundreds of Wakamba dead by the roadside are not only horrific and pitiable—they are graphic testimony to the prey available for scavenging predators. Things were only marginally better for the Taita people, as determined by Field Museum anthropologist Chap Kusimba, who gave me access to his article on "Early European accounts of the peoples and cultures of Tsavo." The Wataita had experienced crushing droughts in the 1880s that had caused extensive migrations of people to Taveta, Ukambani, and elsewhere. They had even sold their children into slavery in exchange for food. Their range in Tsavo contracted to perhaps a fifth or a sixth of its former extent as a result of these droughts and famine.

The chain of causation for man-eating from scavenging human corpses is substantially shorter than that from dead cattle—it involves solely a shift from eating dead people to eating living ones. The drought initiated the famine that began in mid-1897, so that drought may also have been a trigger for man-eating at Tsavo. Drought is also known to speed the transmission of rinderpest—by concentrating the drinking activities of susceptible hosts at a handful of pools, healthy animals are more rapidly exposed to the infectious saliva and feces of sick hosts (Dobson 1995).

Caravans Left Dead Bodies for Scavengers

Hardy believed that caravans were the key to why the Tsavo lions ate railroad workers. In *The Iron Snake,* he stated, "There is no accepted explanation of the continuous presence of man-eaters at Tsavo. Some claim (without biological justification) that the soil and water of Tsavo produced a salt deficiency in the lion's natural prey and that the lions compensated themselves in human flesh and blood. Probably the root lies in the trade and slave caravans that used the Tsavo path and encamped by its river.

Disease was never far and there were always losses. Bodies were certainly left to the scavengers" (1965, 314). Although Hardy's book may well be fictional, it is well researched, and the scenario he paints is highly plausible.

As documented elsewhere, opportunistic predators may first encounter specific prey by scavenging their remains. This frequently happens in the wake of regional calamities, such as a disease epidemic, in which the scale of mortality exceeds the abilities of human survivors to properly dispose of their dead. By leaving human remains where lions, tigers, and leopards can scavenge them, people alter the availability of humans in the carnivores' diets, tragically setting the stage for man-eating behavior, as Kenneth Anderson (1955) and others have noted.

Unlike the case in other regions of Africa, East Africa's slave trade was inextricably linked to ivory. This story is gradually coming into focus with new research by Rahul Oka, Chap Kusimba, and Peter F. Thorbahn, again made available to me by Dr. Kusimba. Before recorded history, Arab or Swahili traders had headed into Africa's hinterland in search of ivory; rhinoceros horns, cat skins, rock crystals, iron bloom, and other goods were also valued trade objects. Any ivory hunted or purchased would have to be carried out, all the way to the coast, and some of the tusks in precolonial days were extraordinarily large, weighing 100 to 175 pounds. Because beasts of burden fared poorly in tsetse fly country, traders routinely rounded up slaves on their safaris, knowing that the odds were good that their service as porters would be needed at some point on the itinerary. Even if the trip failed to collect an impressive quantity of ivory, the slaves could always be sold at retail (as "black ivory") back on the coast, to cover the costs of the safari.

Zoologist Clive Spinage reviewed the history of ivory trade in East Africa (1973). Caravan traffic in the area around Tsavo grew substantially in the nineteenth century, along with Arab interest in Zanzibar. During this period, Mombasa was exporting one hundred thousand pounds of ivory per year, and Tanga and Pangani were annually exporting seventy thousand pounds and thirty-five thousand pounds, respectively. In 1850, Mombasa's annual ivory exports represented 1,650 porter loads, so that the total size of caravans supplying these resources might have been three thousand to five thousand people.* All of these people would have passed through the greater Tsavo ecosystem, either via the Taita-Taveta route used by Thomson to reach Masailand or across the Taru Desert into Tsavo itself. Caravans plying the Sabaki route blazed by Lugard carried goods into Malindi, which

*These numbers include traders, guards, and porters carrying food, water, and trade goods. Even Thomson's initial survey expedition consisted of more than a hundred men.

was itself a major shipping port, as Oka and colleagues have shown.

After his time in Tsavo, Col. Patterson witnessed one of these caravans for himself. While hunting on the Athi Plains, which are bisected by the railroad southeast of Nairobi, his party was passed by a great caravan. The multitude included some four thousand men, Basoga porters on their way to the coast carrying goods for a Sikh regiment returning from Sudan and Uganda. On the coast, many fell victim to dysentery, and the group reappeared in Athi in terrible shape. "The ranks of the caravans were terribly decimated, and dozens of men were left dead and dying along the roadside after each march. It was a case of survival of the fittest, as of course it was quite impossible for the whole caravan to halt in the wilderness where neither food nor water was to be had" (Patterson 1907, 215). These conditions prevailed among government-paid porters—imagine how much worse were the conditions for slaves!

The slave trade was not the sole means by which humans have invited man-eating incidents by littering the landscape with human bodies. Then as now, internecine warfare provided ample opportunities. Conflicts between the Maasai and their neighbors, especially the Kamba and Kikuyu tribes, provided a regular supply of bodies in historic times, but such conflicts are not only historic. A former ranger in the Kenya Wildlife Service, Walter Njuguma of Nakuru, gave me a first-hand account of his years on the Somali frontier in the 1970s. Civil war led to bloody excesses in which entire villages were massacred and the bodies of victims were left lying where they fell. Inevitably, vultures, jackals, hyenas, and lions served as undertakers, and the regularity of bloodshed made scavenging human carcasses a common affair. During his service as an animal-control officer, Njuguma shot hundreds of hyenas and lions that had become emboldened by this habit and attacked or threatened persons living in the area. On one occasion, he witnessed a lioness and her cubs approaching a small occupied hut; at the time, he was convinced that the lioness was teaching her cubs the art of hunting hut-bound humans, and dispatched them before they entered the hut. That particular group became part of Njuguma's career total. Nevertheless, the ubiquity of tribal strife and associated warfare in Africa routinely exposes predators and scavengers to human meals beyond the protection of government guardians.

Funeral Rituals Invite Preying on Human Corpses

Colonel Patterson . . . accounted for these young, vigorous animals becoming man-eaters because some of the coolie workers who died were

> *put into the bush unburied and the lions had acquired a taste for human flesh by eating these bodies. After this taste was acquired these lions hunted men just as the ordinary lion hunted zebras. They made a regular business of it.* (Carl Akeley, *In Brightest Africa* 1920, 61)

The railway camps were multinational affairs, making up a polyglot mosaic that existed for the sole purpose of pushing the railroad ever deeper into Africa. Crews had been assembled from various locations on the Indian subcontinent, in Southwest Asia, along the Swahili Coast, and where possible from local villages. Each of these groups brought their own beliefs, customs, and traditions to the camps. This cultural diversity, still evident in Kenya today, was also recognized by the Railway Commission, as detailed in Hill's history *Permanent Way.* It preferentially deployed workers from northern India and Pakistan—mainly from upland regions—to work in the Kenyan highlands and assigned those from sweltering Madras (now Chennai) and other southern regions to coastal duty. These people were thought to be "pre-adapted" to the climate conditions obtaining in these respective zones.

But there was no place left in Western India or Asia Minor where people were accustomed to living alongside large carnivores. Even a century ago, lions had been extirpated from much of their natural range in Asia; while Patterson was in Tsavo, they persisted in India solely in one rajah's reserve (Chellam 1996). Actively predatory hyenas had disappeared from that region at end of the Pleistocene era.

The various attitudes of these people toward the sick and dead are relevant to the events at Tsavo. Along the line, railway laborers were stricken with ulcers, diarrhea, dysentery, liver complaints, malaria, scurvy, and burrowing fleas. By the time Ungan Singh was eaten in Tsavo in March 1898, nearly 5 percent of the 7,131 Indians employed by the railway had died and almost 10 percent of the remainder had become so weakened by disease that they were considered invalids and returned home. The final report of the Uganda Railway Committee noted that of 31,983 Indians employed by the railroad, 6,454 (20 percent) were invalided and 2,493 others (7.8 percent) had died in East Africa.

Final observances for these men rested in the hands of their countrymen, and the procedures followed were a matter of their respective beliefs. As Frederick Selous noted, "No matter how plentiful game may be, lions will almost invariably feast upon any dead animal left by a hunter, from a buffalo to a steinbuck, that they happen to come across" (1881, 265). Improper disposal of human remains could have triggered habitual man-eating among opportunistic lions.

Bombay was a major source of railway workers. Then as now (rechristened Mumbai), it was a haven for Zoroastrians, or Parsis. Adherents of the Zoroastrian faith dispose of their dead by leaving the corpse exposed to the elements in a deep pit (*dokhma*) constructed within a larger structure (the "Tower of Silence"). In the sanctity of this well, the body is typically consumed by vultures. The funeral provisions that Parsis at Tsavo made for their fallen countrymen are unrecorded.

On the other hand, Hindus generally cremate their dead. The body is not simply placed on a pyre and burned. Instead, the deceased is bathed and freshly dressed. The corpse is anointed with fragrant sandalwood paste and then decorated with flowers and garlands. Gold dust is sometimes sprinkled on the face. After scriptural recitation and other rituals, the body is placed on a pyre oriented north to south. A family member circles the pyre while praying. Finally, the funeral pyre is lit after the mouth of the deceased is brushed with kindling.*

This ritual-filled ceremony is impossible when fuel is short, labor is unavailable, families are absent, or victims die in very rapid succession (as during epidemics). At such times, this rite may be waived, or rather performed in symbolic abbreviation. A live coal is placed in the mouth of the deceased in a village ceremony, and then the body is carried out of town and discarded. In his introduction to the 1986 reprint of Patterson's classic book, Peter Capstick guessed that scavenging on fever victims treated in this manner might have initiated the lions' attacks in Tsavo.

Sikhs were also well represented in the work crews, and the horror of Ungan Singh's death helped to initiate the lions' "Reign of Terror." Sikhism imposes no restrictions on the treatment of a body after death. The corpse may be buried, cast into water, cremated, or treated in some other convenient manner, so long as it is respectful.† Convenience often leads Sikhs to cremate the dead, and given that some form of this practice is obligatory for Hindus in the work crews, it seems likely some form of cremation was performed over Sikhs who died during the construction of the railway.

Although few Africans worked on the Railway, there were Swahilis among the work crews. Many were recruited during 1897, when the immigration of Indian workers was interrupted by an outbreak of plague on the subcontinent. However, neither employment prospects, a severe famine engendered by drought (1897), nor rinderpest (1898) led many in the

*http://www.hindunet.org/last_rites/

†http://members.pgonline.com/~mpurewal/chapter02.html

Kamba and Kikuyu tribes to seek railway jobs. Those who sought railway employment typically served as porters, carrying goods to railhead. The conditions under which they camped are unknown but might have resembled those of traders in caravans. Because Swahilis are Muslims, burial of dead bodies is mandatory and cremation is forbidden. Scavenging from Swahili corpses is least likely to have been the cause of the initiation of the lions' attacks at Tsavo.

Various theorists have cited "ritual invitation" (i.e., scavenging on human remains made available through religious ritual) as a possible explanation for the man-eating incidents at Tsavo. The high mortality of railway workers and the exotic diversity of their respective customs demand consideration. Yet, of these diverse means of treating the dead, only the Zoroastrian and abbreviated Hindu (and perhaps Sikh) ceremonies seem conducive to the development of man-eating behavior. And there are no details available on the number of adherents of these religions in the railhead crews or on the number who died in Tsavo before Patterson's arrival, possibly triggering man-eating behavior.

Curiously, none of the authors citing this mechanism have discussed the burial practices of resident African peoples. In fact, the Maasai *rely on scavengers* to dispose of their dead. Believing that burial is harmful to the soil, the Maasai reserve the custom of burial solely for great chiefs. After death, ordinary people are simply carried into the bush at the edge of town and left for scavengers. As discussed under "Opportunity," these customs allow predators to become familiar with the smell and taste of people and inured to the presence of people on their foraging routes. Given centuries of occupation of the area by the Maasai and Kamba peoples, Tsavo's predators regularly confronted corpses that might have attracted their attention, which was later led to railway workers. However, this effect should have been felt regionwide, not localized at points along the railway line.

Naïveté Led to Inadequate Protection from Lions

The camp at Tsavo was filled almost entirely with immigrants having little exposure to East African wildlife and its dangers. The initial naïveté of the workers at Tsavo is starkly shown by Patterson's photo of the tent from which Ungan Singh was taken (Patterson 1907, 23). Taken during the first month of Patterson's residence in Tsavo Camp, it shows a broad expanse of open country dotted with tents. None of these tents has a *boma,* and many lie within meters of the railroad that may well have served the cats as game trails by which to approach the camp.

Patterson later adopted the local custom of constructing thorn *bomas* around all habitations for man and livestock alike. The *boma* constructed around his own compound on the eastern side of the Tsavo River measured some seventy yards in diameter! Col. Patterson also directed *bomas* to be built around the workmen's camps and specified that fires be kept burning in them all night. These precautions proved at times ineffective in curtailing the lions' attacks. However, it could be argued that the initial lack of *bomas* permitted the cats to become accustomed to hunting in the camps and to humans as prey. Once the lions had developed their predilection for humans, no *boma* was thick or high enough to repel them.

Galla herdsmen in Tsavo today continue to use *bomas* to protect their flocks of cattle and goats. Curiously, the cut branches of acacias and wait-a-bits are regularly oriented outward, while the formidable thorns are directed inward, toward the livestock. Such herdsmen laugh at the proposition of constructing and maintaining lion-proof *bomas* large enough to protect their cattle. Instead, they build *bomas* to keep the cows tightly grouped, and protect the *bomas* from lions themselves—by using firebrands and spears, beating pots, and shouting. Watchmen with fires might have proved more effective than thorns and firearms in the railroad camp at Tsavo, had the railway workers employed time-honored African traditions of defense.

Attacks by "Lion Men"

> *In both the later precolonial and colonial periods, there is sufficient evidence to strongly speculate about terrorist killing as a tactic to disrupt the processes of colonial conquest and later hegemony.* (Allen F. Roberts, personal communication, June 8, 2001*)

In the 1890s, along the eastern shores of Lake Tanganyika, "lions" killed scores of Tabwa people living around the White Fathers mission at Badouinville. All the victims had been prospective converts to Christianity. Survivors begged the priests to banish the supposed "lion-men," people who assumed the shapes of lions to rampage through the community. The

*I am greatly indebted to Professor Roberts, director of the African Studies Center at UCLA, for raising the possibility that the attacks at Tsavo might have a human cause and for guiding my investigation into relevant literature, much of it authored by him.

priests simply exploited the superstition of the people and used the threat of attacks to restrain locals from dispersing. The attacks continued. Deaths continued episodically until the 1930s, when fifty Tabwa died, and the colonial administration finally recognized the murderers as human. The perpetrators were hanged and the story there closed, but "lion men" continued to appear and operate as terrorists in various parts of Africa over the following half-century.

The "lion man" phenomenon is not totemic. Although some African societies practice totem worship, this practice rarely if ever reaches the prominence it has among Native Americans. In southern Africa, a form of totemism existed among the Bechuanas, where each clan held particular animals as sacred and abstained from eating their meat, in some cases even praying and making sacrifices to them. Members of these clans became known as "buffalo-men," "crocodile-men," "leopard-men," and so on. But most Bantu tribes do not maintain totems for worship (Nassau 1904). Among the Kaonde, who inhabited the Kasempa District in northern Congo, people believe in reincarnation. This transmigration of souls fills the bodies of the living with the souls of the dead; some, awaiting a new body in which to reside, rain hardship, sickness and pain on people opposed to their interests. Most spirits reenter people, but if a chief dies having taken special medications prior to his death, he may return as a lion (Melland 1923). However, unlike these clansmen or former chiefs, the "lion-men" or *watuSimba* of Tunduru were hardly religious—instead, they were instruments of terrorism and murder.

Another group of theriomorphs (literally "beast-forms"), the "lion-men of Singida," were actually women (Schneider 1982). Captured and enslaved as girls, they were locked in cages and plied with drugs by witch doctors until they became insensible and compliant. They were then taught to walk on all fours, sewn into animal hides, and fitted with deadly prosthetic claws—their purpose was to be sent out at night on assignment as assassins (Cowie 1966). The cult persisted for decades in the Singida District of Tanzania, sustained by the fear and superstition of rural people. In 1961, the regional commissioner of Tanzania's African National Party called for their extinction: "Those still posing as lions and walking on all fours and killing others must be routed out. We must bury the traditions of the past" (Blau 1961).

"Lion-men" continue to influence rural actions even today. So gripping were their reigns of terror in the early twentieth century that even an old hand like Rushby shot numerous lions mistakenly believing them responsible for the murderous rampages underway in his district. More recently, Marcus Borner, a biologist with the Frankfurt Zoological Society, reported that more than 250 lions had been killed in southern Tanzania over a two-year period in connection with human deaths. But it was not known how many of the slain lions were actually man-killers, because some of the deaths attributed to lions in Tunduru may have been paid murders by "lion-men."*

In most of these instances, "lion-men" are a political or economic phenomenon, not religious. As Professor Allen Roberts, now at UCLA, put it, "Attacks by lion-men are a political strategy of terrorism" (1986, 69). The outbreak of "lion-men" among the Tabwa at the turn of the last century coincided with and was a direct response to colonial expansion in Congo. The "lion-men" became the disguise adopted by native guerrillas to thwart European expansion and counter European interests. A Christian mission became the focus of these attacks in Congo; what better symbol of European expansionism in Kenya than the railroad? How better to safeguard the ancestral home of Kenyans than to sabotage the railroad's construction? The attacks struck fear into the hearts of railway workers and fostered dissent, discord, and desertion. They certainly slowed the construction effort.

Roberts might have been the first to hypothesize that the Tsavo incident was the product of "lion-men," but the idea predated his formal articulation. In *The Iron Snake*, Ronald Hardy wrote that the killings in Tsavo Camp might have been a combination of real lions and villains who disguised their crimes by masquerading as lions: "While Patterson was perched in a tree over Singh's tent, waiting for a shot, a heartrending shriek came through the night. Preston and Turk ran up to the victim's tent: The canvas hung, ripped. Outside, the marks of the drag filled with rain. The hedge of thorn was broken and, bending, they saw the rags of cotton from the coolie's dress, a turban, all heavy with rain and, the flesh trembling in small brown morsels on the spines of thorn" (1965, 130). The next day, they discovered that the body of the man had been dragged a hundred yards from the tent into a ravine and had been three-quarters eaten.

On reporting to Whitehouse, they discovered another man had been taken that night, a *jemadar,* but on examining the body found that, although his jugular was slashed and his chest deeply clawed, there were only sandal tracks on the ground where he lay, evidently the victim of murder. "Something was wrong. The claw-marks was wrong, as if some crazed thing had gone mad at him, it might have been Leopard but there were no pugs and not a meal taken from him.... I said nothing to W. He is worried enough...." (Hardy 1965, 131).

It is remotely possible that the *Wakamba* exploited a rash of lion

*http://lynx.uio.no/catfolk/cnissues/cn12-15.htm

attacks and turned them into a veritable "Reign of Terror" in order to slow imperial expansion. This idea gains credence when one considers the peripheral, sometimes adversarial role of Africans in the railway. Aside from Swahilis, who were nearly as foreign and threatening to the *Wakamba* and *Wataita* residents as were Sikhs and Hindus, few Africans were employed by the railway. The roadbed undoubtedly crossed village boundaries, desecrated burial grounds, appropriated prime water sources, and otherwise ran roughshod over the belief and value systems of native Africans. Further grounds for resentment might have arisen from colonial suppression of tribal traditions, such as raiding parties, slavery, and religious practices. Obviously, the British clashed with local tribes over appropriate responses to the rinderpest outbreak: The *Wakamba* disregarded the precautions mandated by Sub-commissioner Ainsworth, and the *Wakikuyu* initially ignored and subsequently murdered Veterinary-Captain Officer Haslam.

However, the material evidence that actual lions ravaged the work crews at Tsavo refutes the "African guerrilla" hypothesis. Some of the attacks on the railway camp were entirely inferential—informants saw only remains of a corpse in the bush, or heard only a man's piercing cries of distress immediately prior to violent death. Superstition, context, and mass hysteria could certainly fill in many of the blanks. But as the "Reign of Terror" continued and the lions became more brazen in their attacks, they were repeatedly witnessed attacking men, carrying them off, and even consuming them within sight of the camp. And the cultural multiplicity of the work crews—their ethnic and linguistic diversity and the accompanying division of the workforce into well-defined occupational guilds—would have inhibited the spread of mass hysteria. Finally, if this hypothesis were true, then Patterson's shooting of the two lions would not have ended the attacks. Instead, for at least for a time, lion attacks on people near Tsavo ended in December 1898.

The Likelihood of Multiple Causes of Man-eating Behavior

From the foregoing review, it should be obvious that no single hypothesis can account for the diverse scenarios in which man-eating behavior is expressed. Different hypotheses offer cogent explanations in different scenarios—none is universally applicable. Moreover, some of the hypotheses share predictions and mechanisms, leaving an investigator of any specific instance of man-eating with a set of overlapping possibilities rather than testable hypotheses with unique predictions.

Therefore, it is seldom possible to trace causation to a single hypothesis. Man-eating is a complex behavior pattern that is learned, not innate. The likelihood of man-eating in any instance depends on the age, condition, and prior experience of the lion, as well as the traditional and current habits of the people involved—their settlement patterns and ways of living. It is also affected by the current environmental conditions and recent history of the area. With so many variables, generating concrete predictions is virtually impossible. And as touched upon above, human superstition clouds the subject of man-eating. It impedes getting a clear reading of fact, making it impossible to determine cause and effect.

Lion attacks in the Tunduru District of Tanzania offer a compelling example of multiplicity.* In one attack, a local game warden was killed in the midst of six hunters. In another, a lion chased a dog into a house, where it seized a 10-year-old boy from a sleeping group and remained in the house eating him until dawn, when the boy's father shot it. According to *Cat News*, these attacks may have been triggered by depletion of wildlife ("prey shortage"), because poaching had wiped out most of the large animals in the area. Lion predation on cattle and goats became so high that animal husbandry became unprofitable, and local people turned to harvesting bushmeat as a protein source ("familiarity" and "injury"). With people in the bush seeking bushmeat, predators crippled by snares, and lions patrolling settled areas looking for domestic stock, encounter rates between lions and humans skyrocketed ("opportunity" and "familiarity"). The lions may also have acquired a taste for human flesh from eating the bodies of victims of the bush war that ravaged neighboring Mozambique ("human invitation"). How could these episodes be ascribed to any single cause?

*http://lynx.uio.no/catfolk/cnissues/cn12-15.htm

Chapter 5
Lion Biology: Evolution and Geographic Distribution

To understand lions, one needs to understand their evolution and distribution, their ecology and behavior. All organisms are the products of extended evolutionary histories, and many of their characteristics are part of their genealogical endowments, "evolutionary baggage," as it were. These traits reflect the environments and exigencies of their ancestors but some may be maladaptive or irrelevant in their present biological and environmental contexts. Viewing the characteristics of lions alongside those of other cats allows us to specify which are unique to lions, what are their likely adaptations, and which represent leftovers from a prototypical cat ancestor.

Evolution of Cats

Felids, members of the cat family Felidae, first appeared in the Old World, with the appearance of *Proailurus*, a bobcat-sized cat, at the beginning of the Miocene (about twenty-four million years ago). This animal was soon followed by *Pseudaelurus*, which appeared in Asia twenty million years ago and reached North America two million years later. Throughout the Miocene, felids in both Eurasia and North America increased in both diversity and body size. By the Miocene-Pliocene boundary (five to eight million years ago), essentially modern guilds of large felids, wolflike canids and bone-cracking hyenas had evolved (Van Valkenburgh 1999). The appearance of essentially modern faunas may have been triggered by replacement of moist woodlands in the Mediterranean region by seasonally arid grasslands (Janis 1993). However, these faunas were even richer in adaptive forms. No modern felids have saber teeth, but members of several cat lineages exploited this strategy in the geologic past. The disappearance of saber-toothed cats resulted from competition with species of the newly evolved genus *Panthera* (which today includes lions, tigers, leopards and jaguars), as well as from their dependence on very large herbivores that went extinct at the end of the Pleistocene.

Living cats have long been divided by their larynges, or voice boxes, into "roaring" and "purring" groups (Hast 1986; Peters & Hast 1994). It is true that the great roaring cats can generate a purring sound, but this vocalization differs from that of purring cats in that sound is produced only on exhalation rather than on both inhalation and exhalation (Packer & West 2000). The divergence, or evolutionary split, of the two groups apparently took place near the Miocene-Pliocene transition. Although representing only one of many characters modified in the course of the group's evolutionary history, larynx type and roaring vs. purring identify a fundamental evolutionary dichotomy in cats—the structural distinctions of these groups have been recently affirmed and detailed by Chicago anatomist Malcolm Hast and German zoologist Gustav Peters (1994).

The evolutionary relationships of cats have been recently analyzed by two Canadian biologists, Michelle Mattern and Deborah McLennan, using molecular characters (DNA sequences), morphological characters (mainly skull, dental, and skin characters), and chromosomal complements (2000). This "total evidence" approach to reconstructing relationships among cats employs all available evidence. It typically offers better resolution of taxonomic relationships and clearer statements of character evolution than reconstructions relying on only a single set of traits (Flynn & Nedbal 1998). Mattern and McLennan's total evidence analysis united all the big cats near the root of their evolutionary tree. This indicates that the roaring cats are the basal-most (earliest-originating) group of living cats, a "sister" to (having its closest genealogical relationship with) all remaining cats. This interpretation isn't new—it supports British zoologist R. I. Pocock's proposal in 1917 that the roaring cats constitute a distinct subfamily of Felidae, the Pantherinae.

Dating the evolutionary split is a bit trickier, as it is strongly dependent on assumptions about rates at which DNA evolves. Using the rate of nucleotide base-pair changes (substitutions) over short DNA sequences as a "molecular clock," molecular biologists Warren Johnson and Steve O'Brien estimated the *Panthera* clade (or branch of the cat tree) to be 6.0 million years old (1997). However, their tree for the Felidae was based solely on substitutions from two short nucleotide segments, which indicated that the *Puma*, or mountain lion, lineage had diverged at least 8.2 million years ago. But because pantherines are now thought to be basal to the appearance of the *Puma* lineage, the big cats must have appeared earlier, sometime in the late Miocene.

Despite the early origin of the pantherine group, little genetic divergence separates the living species. Some see this as evidence that the modern species originated during the last two million years. For example, in 1996, Warren Johnson and colleagues stated, "Species of the genus *Panthera* ... evolved recently and rapidly" (1996). However, accumulating fossil evidence indicates that modern species of roaring cats appeared earlier and that the DNA substitution rate has been slower than expected.* South African paleontologist Alan Turner argued that the living African pantherines, lions and leopards, date back to the mid-Pliocene, about four million years ago (Turner & Antón 1997).

It is impossible to specify the geographic origins of a group without an accompanying hypothesis of evolutionary relationships (phylogeny). Without a phylogeny, time of first appearance in the fossil record is often used. The oldest known *Panthera* species hail from the Pliocene in Asia, represented by the tigerlike *P. palaeosinensis* in northern China and the jaguarlike *P. gombazogensis* in northern Eurasia. Their age and distribution has been interpreted to mean that all the pantherines must have evolved in Asia, or at least have their ancestors there.

The branching structure of a cladogram (genealogical tree) specifies the precise sequence of appearance of each lineage in the group's evolution. The same structure can also be used to reconstruct and specify geographic ranges. The analysis of Mattern and McLennan (2000) identified tigers and jaguars as the most highly derived living pantherines and lions as their sister group. If *P. palaeosinensis* and *P. gombazogensis* are in fact relatives of tigers and jaguars, then their Asian distribution furnishes no clues to the origins and affinities of lions—only lower, more basal branches can tell us where lions and their ancestors lived. The tree's structure implies that tigers and jaguars had a fossil progenitor, whose distribution is unknown but probably Asian, and that this common ancestor evolved *after* the divergence of lions from other great cats.*† Because tigers, jaguars, and their fossil antecedents evolved subsequently, their distribution is uninformative about the origins of lions. The lineage containing lions, tigers, and jaguars has as successively distant relatives leopards and snow leopards (see Figure 8).

*Fossils are used to "calibrate" the molecular clock, to account for differences between groups in rates of nucleotide substitution. Accordingly, independently established fossil dates can be used to invalidate the dates of divergences as estimated by the molecular clock, but the reverse is not true. One can question the accuracy of fossil identifications, however.

†This reasoning follows the *progression rule*, first formulated by the German systematist Willi Hennig. The distribution of an ancestral species is inferred from the ranges of its descendants—one has only to work backward through the tree, assigning inferred distributions to hypothetical common ancestors, represented by each bifurcation point (node) that unites branches. If two descendant species or lineages have nonoverlapping ranges, then the ancestor is inferred to have lived throughout the ranges of both. If the ranges of descendants overlap, then only the area of overlap is the presumed range of the ancestor (Hennig 1996). By the topology of Fig. 8, the common ancestor of lions and its sisters lived in Africa and Asia.

Unfortunately, this set of relationships sheds no light on the geographic origination of lions or leopards because both occur and are (or were) widespread in both Africa and Asia. Either continent could be their ancestral homeland.

However, we should note in passing that the cladogram does help interpret the evolutionary development of various morphological and ecological traits of pantherines, especially lions. The tree's structure identifies tufted tails and the manes of male lions as uniquely derived character states found only in lions. These unique traits likely appeared after lions diverged from the other great cats, so that the common ancestor of lions, tigers, and jaguars probably lacked these characteristics, that is, was maneless and tuftless. "Large body size" is an ecological trait that lions, tigers, and jaguars share to the exclusion of their sister group, leopards, and presumably typified the lion ancestor. "Diurnal habits" and "savanna associations" are supposedly unique to the lion lineage among the great cats (Mattern & McLennan 2000).

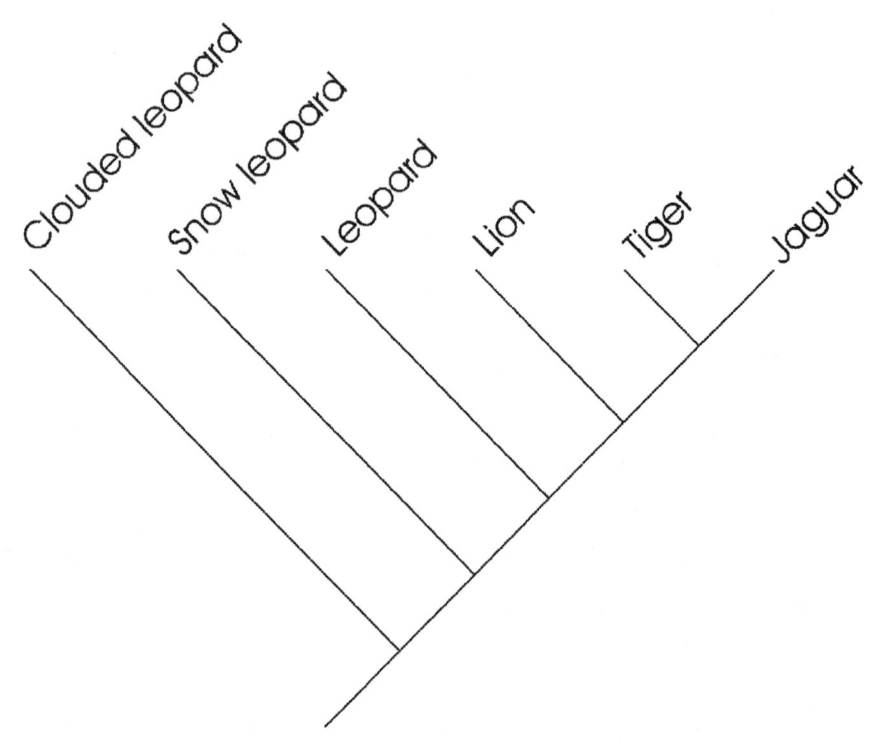

Figure 8. A cladogram or branching diagram of evolutionary relationships among pantherine cats, as determined by a "total evidence" analysis by Mattern and McLennan (2000).

Another obviously derived character of lions is their uniform coloration, a distinctive pelage trait amidst branches of spotted and striped great cats. This trait evolved independently in jaguarundis and pumas, among others, in the purring group. A recent phylogenetic analysis of coat color evolution by Swedish zoologists Lars Werdelin and Lennart Olsson (1997) identified a "flecked" pattern (i.e., simple spots, as in the cheetah) as the ancestral state for cats. All other coloration patterns (rosettes, vertical stripes, small blotches, large blotches, and uniform coloration) can be derived from this one by relatively subtle changes in such parameters as body size at the onset of pigment deposition. It is significant that lions are born completely flecked and retain flecks and rosettes on their bellies (sometimes with traces on their limbs) into adulthood. This coloration pattern is a persistent indication of their evolutionary legacy.

Alan Turner's studies of fossil remains have probably shed the most light on the place and time of origin for lions. He tallied the presence of various extinct and living carnivores among fossil-bearing sites in Africa since the late Miocene (1985). His table of first and last appearances of species at these sites over the last nine million years offers a "deep time" perspective on faunal history, one that emphasizes species turnover. Individual species appear and disappear, often for idiosyncratic reasons, but wholesale shifts in assemblages are usually thought to indicate major environmental shifts. Roughly four million years ago, Africa's carnivore fauna experienced a dramatic changing of the guard. Various archaic forms were replaced by essentially modern species. Both cheetahs and leopards first appear in the fossil record in Africa between three and four million years ago, and both have modern distributions in Africa, Asia Minor, and southern Asia. Although leopards are catholic in terms of preferred habitat, the appearance of cheetahs would seem to signal the expansion and widespread distribution of open grasslands.

When lions first appeared is still uncertain, the date hinging on the identity of the single *Panthera* jaw bone from Laetoli. Alan Turner identified the Laetoli jaw as indistinguishable from that of modern *P. leo,* arguing that there is no evidence to identify it as coming from anything but a lion. It had earlier been assigned to either *P. palaeosinensis* or *P. gombazogensis,* but the Laetoli jaw is demonstrably larger than that of either Asian species (Turner & Antón, 1997).

Unequivocal fossils of *P. leo* appeared later, between two and three million years ago,* and these records blanket eastern and southern Africa. Lions

*Ca. 2.3–1.9 million years ago in East Africa and 2.8–2.4 million years ago in South Africa (Turner 1985). A number of these records predate the earliest records noted by Nancy Neff (Coheleach 1982). Neff's dates were used in early applications of the molecular clock to the first appearance of *Panthera.*

first appeared in Europe in southern France (at Vallonet) around nine hundred thousand years ago and are a common element of European Ice Age faunas thereafter. The appearance of lions in Europe at this time (the Lower-Middle Pleistocene boundary) is remarkable, for it roughly coincides with the first appearance of hominids in the temperate zone, dispersing from the tropics. Lions, leopards, spotted hyenas, and humans all moved in parallel at roughly the same time, suggesting a common environmental trigger. There is strong evidence in the fossil record that there were repeated dispersal events—anywhere from three to seven—of human populations in southern and eastern Africa (Strait & Wood 1999).

During the Pleistocene, climate was continually changing, which shuffled and redistributed the vast mosaic of the African savannas. These changes are reflected in the faunal composition of Lainyamok, a mid-Pleistocene site in Kenya studied by Smithsonian paleobiologist Rick Potts and colleagues. There, both geographic ranges and taxonomic associations of large-bodied mammals fluctuated extensively over the last three hundred thousand years, sweeping over much of eastern and southern Africa (Potts & Deino 1995). As a result of this continual mixing, eastern and southern Africa share fundamental similarities in their lion populations. In some cases, often involving smaller organisms, modern species have "disjunct," or interrupted, distributions, occurring in both eastern and southern Africa but not in between. However, larger organisms have greater dispersal powers and more continuous distributions. Lions may have ranged more or less continuously through the arid savannas that blanket eastern and southern Africa until their recent range collapse from both persecution and habitat fragmentation.

If both lions and leopards originated and evolved in Africa prior to invading Asia, then some of the earliest divergences in each line would occur among African populations. This is the pattern seen in many studies of human genetic variation (Excoffier & Schneider 1999). However, neither cat fulfills this prediction. Studies of genetic variation in leopards by Sriyanie Miththapala and colleagues at the National Cancer Center show relatively deep divergence between African and Asian populations, but populations in western Asia, eastern Asia, and Indonesia show far greater differentiation among them than do regional samples in Africa (1996). Comparable genetic surveys are lacking for the entire range of the lion, but surveys of eastern and southern African populations by geneticist Jean Dubach of Chicago's Brookfield Zoo and colleagues show only modest genetic variation. Until West African and Indian populations are added, those results will be incomplete and necessarily inconclusive. However, the genetic uniformity of both lions and leopards in Africa argues either for their origin and evolution elsewhere, probably in Asia, or for severe population "bottlenecks," which collapsed genetic diversity to that in a few surviving lineages, in Africa.

Size Variation of Lions

As indicated earlier, lions first appeared in Europe in southern France about nine hundred thousand years ago.* Paleontologists have assigned their remains to various species, including tigers, but only size distinguishes them from modern lions. The older lions, usually classified as *P. leo fossilis*, were up to 25 percent larger than modern lions or tigers in length of skull, size of teeth, and length and robustness of various limb elements (Ballesio 1975). Over the last million years, the average size of European lions decreased, until in the cave lion, usually designated *P. leo spelaea*, it was only 8–10 percent larger than that of modern lions. Because of body size variation associated with age, sex, and individual differences, it is very difficult to specify the size of any population from one or a few individuals. This probably accounts for the various scientific names applied to European lions. Although Pleistocene lions were larger on average than modern ones, there was substantial overlap in size between most modern African lion populations and late Ice Age populations of *P. leo spelaea* (Turner & Antón 1997).

During the Pleistocene, lions also occurred in the Western Hemisphere. Lions are known from more than forty North American localities and ranged as far south as northern Peru! (Kurtén & Anderson 1980). Usually distinguished from contemporaneous lions of Eurasia as *P. atrox*, the North American lion rivaled *P. leo fossilis* in size but had a larger brain. However, in view of their wide ranging habits and great dispersal ability, it now seems likely that the forms *atrox* and *spelaea* were regional varieties (subspecies) of a single biological species. Brown bears, wolverines, and red foxes represent better-studied modern-day analogues—historically, Eurasian and North American members of these species were also regarded as distinct species.

Most estimates of body mass in the literature derive from a simple scaling relationship based on width-to-length ratios of bones. Because body weight increases as the cube of body length, the bones of progressively larger animals must be stouter in shape to support their substantially heavier weights. However, increases in diameter are not the only way to strengthen

*There are prior records of large cats in Europe, the so-called Tuscany lions, but these were probably more closely related to leopards (Guggisberg 1975).

the marrow-bearing long bones of the limbs; thickening of the bone's cortex (wall) can also increase its strength without altering its external dimensions. In an ingenious survey of modern carnivores while studying at UCLA, Kenyan paleontologist William Anyonge examined the relationship between mass and cross-sectional properties of some carnivore long bones, particularly the ratio of cortical bone to medullary (marrow) cavity size. He then studied bone thickness among fossils. His estimates of mass for two of the extinct carnivores represented in the La Brea "tar pits" of Los Angeles—the sabretooth cat *Smilodon fatalis* and the lion *P. atrox*—were up to one and a half times heavier than previously believed. Anyonge estimated that the American lion *P. atrox* weighed 750–1150 pounds, about twice the weight of modern lions and comparable in size to the largest living bears (1993). Substantial size variation in brown bears, ranging up to the enormous Kodiak Island bear, proves that this size variation need not signify distinct species.

Very large size in Pleistocene lions was an evolutionary response to climate and the very large size of their prey, the Pleistocene "megafauna." When lions roamed Eurasia and the Americas, those regions supported mammoths, mastodons, woolly rhinos, various bison, musk oxen, and giant deer. With the sudden extinction of the Pleistocene megafauna from both North and South America about eleven thousand years ago and its concomitant extinction or contraction from Europe and Asia into Africa and India, both the body size and the geographic range of lions shrank accordingly. This range contraction was speeded by climate and habitat shifts. Many Eurasian habitats during the Pleistocene were more open than at present, and much of the continent was covered by a periglacial steppe. The contraction of the lion range coincided with the postglacial expansion of Eurasia's coniferous forests. Both lions and tigers were present in glacial Alaska, and it seems likely the two species had abutting ranges along the forest-steppe interface that stretched all the way from there to Asia Minor.

Geographic Variation and Subspecies of Lions

Just as uncertainty and controversy surrounds the identification and classification of fossil lions, modern populations are subject to differing assessments. Most authorities, including all international regulatory agencies, embrace a classification that includes all extant African lions as members of a single subspecies, *P. leo leo,* and extant Asian lions as *P. leo persicus* (Ellerman & Morrison-Scott 1951, 1966; Meester & Setzer 1971). On the basis of well-marked differentiation, particularly in the size and distribution of the mane, authorities commonly distinguish the extinct populations of Barbary lions and Cape lions from the widespread African lion as distinct and valid subspecies (Guggisberg 1975). However, surprising as it may be, lions have never been subjected to a comprehensive quantitative analysis of either their morphology or their genetics. Consequently, the prevailing view that all existing African lions should be "lumped" into a single highly variable subspecies remains provisional.

Periodically, there are claims that African forms of *P. leo* should be "split" into a number of species or subspecies. A recent proposal, put forward by Chicago zoologists Tom Gnoske and Julian Kerbis Peterhans, involves the lions of Tsavo and has attracted much interest and attention. In two recent articles (Gnoske & Kerbis Peterhans 2000; von Buol 2000), they suggested that Tsavo lions represent a "missing link" between Pleistocene and modern lions on the one hand and Asian and African lions on the other. They designated this lineage as "buffalo lions" and distinguished it from other "pride" lions by its close association and dependence on African buffalo and by certain morphological traits. Until a fuller, more technical argument is published, it is difficult to evaluate their proposals. However, for reasons developed more fully below and in succeeding chapters, Dubach, Kays and I believe that the currently accepted classification is correct: There is a single species of lion in Africa, and Tsavo lions have no evolutionary relationship with either Pleistocene lions or Asian lions that is not shared by other African lions.

Such differing perspectives are not rare in establishing classifications for any group. In each debate, there are "lumpers" (those favoring one or a few recognized names), who tend to be impressed by commonalities among the forms to be classified (or the variability of group-identifying characters), and there are "splitters" (those advocating multiple names), who usually emphasize these distinctions. In a debate over the validity of a "buffalo lion" lineage, my colleagues and I are lumpers, while Gnoske and Kerbis Peterhans are splitters.

Often, there is a bit of arbitrariness to such decisions over taxonomic rank, for example, whether to recognize a group that emerges from analysis as a distinctive taxon and whether to name it as a species or a subspecies. Different views are ultimately arbitrated by the scientific community, which is interested and knowledgeable enough to evaluate the strength of evidence supporting each viewpoint. Scientists ask questions like: Are the methods used widely used, explicitly stated, and repeatable? Are the yardsticks of differentiation consistently applied? Are the standards used in the focal group (in this case, lions) consistent with those employed in related groups (within-species classification in leopards and in tigers, for example)? Only then do other scientists adopt a proposed classification.

Actually, lumpers and splitters have debated the variability and distinctiveness of lion populations throughout the history of the group's classification, and the single-species interpretation is the current verdict. Such prominent authorities as Frederick Selous, James Stevenson-Hamilton, and C. T. Astley Maberly—each with years of field experience involving hundreds of lions in their environments—maintained and emphasized the extraordinary variability of *P. leo.* Each took a dim view of the flurry of names being applied to variants of what they regarded as a single variable species, and flatly rejected the idea that various distinct species of lions inhabited Africa (Astley Maberly 1963; Selous 1881; Stevenson-Hamilton 1912).

But famed explorer Edmund Heller was not exactly an office clerk, and he resolutely believed in the validity of geographic races in lions. A hugely accomplished museum collector, who combed the wilderness of Africa and the Americas for the Field Museum and the Smithsonian before becoming director of the Milwaukee and then the San Francisco zoo, Heller named two subspecies of lions, one of these after Theodore Roosevelt. In justifying the propagation of names for geographic races of lions, Roosevelt and Heller observed in *Life-histories of African Game Animals,* "Much diversity of opinion exists among sportsmen as to the actual existence of the differences assigned by naturalists to the various described races. Some of the hunters who have had the widest experience with lions have observed such great color and pelage variation among them that they refuse to accept the differences which naturalists have pointed out in their diagnoses.... [But t]The racial differences upon which the races of lions are based are so fine that they can only be detected and appreciated on the actual comparison of specimens. Our field observations on differences usually lead us astray."

Such beliefs have produced a dizzying array of names for African lions, which are shown chronologically in Table 5.1. Except for the extinct Cape and Barbary races, all are currently considered synonyms of *P. leo leo*.

The Asian lion (*P. leo persicus*) has been isolated from African lions for thousands of years. Perhaps during the recent period they were connected via populations in the Middle East and northeastern Africa, but it is possible that they have been isolated since the end of the Pleistocene. One molecular

*Decreased frequency of divided foramina over the last 140 years presumably is a result of genetic drift. Fused roots in the lower third premolar is a common trait among Gir lions. The absence of this tooth was first noted in 1910, and the frequency of toothless lions has increased over the ensuing half-century (Todd 1966).

Table 5.1. Formal scientific names proposed for African lions since Carolus Linnaeus (1758).

Taxon or subspecies name	Author, year: page	Country (locality) where name is based
leo	Linnaeus, 1758: 41	Algeria ("Africa"; fixed as Constantine)
barbaricus	von Meyer, 1826: 6	Algeria ("Barbary")
senegalensis	von Meyer, 1826: 6	Senegal (not Felis senegalensis Lesson 1839, a serval)
africanus	Brehn, 1829	Africa
barbarus	Fischer, 1829: 197	Algeria (based on Cuvier's "Lion de Barbarie")
capensis	Fischer, 1829: 565	Based on Griffith's South African Lion (preoccupied)
melanochaitus	Smith, 1842: 177	South Africa (Cape of Good Hope)
gambianus	Gray, 1843: 40	Gambia ("Interior of Gambia")
nubicus	Blainville, 1843: 58, atlas	Egypt (Nubia)
nigra	Loche, 1858: 7	Algeria (Constantine)
somaliensis	Noack, 1891	Somalia ("Somaliland")
kamptzi	Matschie, 1900: 92, fig.	Cameroon (Yoko, upper Sanaga)
massaicus	Neumann, 1900: 550	Tanzania (Kibaya)
sabakiensis	Lönnberg, 1908: 22	Tanzania (Kibonoto [=Kibongoto])
bleyenberghi	Lönnberg, 1913: 273	Democratic Republic of Congo (Katanga)
nyanzae	Heller, 1913: 4	Uganda (Kampala, north shore of Lake Victoria)
roosevelti	Heller, 1913: 2	Ethiopia ("Highlands near Addis Ababa")
azandicus	Allen, 1924: 224, pl. 37–40	Democratic Republic of Congo (Vankerckhovenville)
hollisteri	Allen, 1924: 229	Kenya (Lime Springs, Sotik)
krugeri	Roberts, 1929: 91	South Africa (Kruger National Park, Brixton, Sabi Game Reserve)
vernayi	Roberts, 1948	Botswana (Central Kalahari)
maculatus	Heuvelmans, 1955	Kenya
webbiensis	Zukowsky, 1964	Somalia

study dated their divergence as being about one hundred thousand years before the present, well before the last glaciation, but this analysis is clouded by the extreme population bottleneck experienced by Asian lions a century ago. The sole surviving Asian lion population, which inhabits the Gir Forest in western India, is the final remnant of the race that inhabited Ice Age caves across Eurasia.

The Asian lion is known for its sparse mane and a flap of skin hanging from the abdomen (Chellam 1996). Metrical comparisons of its skull with those of various African lions show it has a shorter palate, broader face, and more constricted braincase. Two nonmetrical characters also typify Gir Forest lions: division of the infraorbital foramen (a cranial passageway below the eyes for nerves and blood vessels) and fused roots of the third lower premolar tooth* (Todd 1966). Both of these traits appear to be random mutations that became fixed in this population owing to "founder effect" (i.e., the population originated from a very small number of individuals, containing a limited sample of the Asian gene pool) and subsequent "genetic drift" (random changes to the composition of the gene pool resulting from which individuals leave behind progeny). The inbred condition of Indian lions is confirmed by their low genetic variability and high incidence of abnormal sperm.

The Cape lion, the first African lion to go extinct, ranged over South Africa's Cape and hinterland. According to Fitzsimons' *Natural History of South Africa*, it was plentiful near Cape Town until 1801, was last reported from the Cape Colony in 1842, and disappeared from Natal in 1865 (1919). It was reputed to be the largest African lion and, like the Barbary lion, characteristically sported an extensive black mane that included a "belly mane" (a midline fringe on the abdomen). Although captive lions sometimes develop very extensive manes, the Czech zoologist Vratislav Mazák showed that Cape lions were distinctive in having a tuft of hairs on each flank of the belly region. Cape lions also had broader and shorter (more bulldog-shaped) skulls than lions to the north did (1975).

The Barbary lion was best known to Europeans and frequently appeared in gladiatorial combat in Rome's Colosseum. This was a well-differentiated race of lions that inhabited the entire Great and Little Atlas mountain system of North Africa (in present-day Morocco and Algeria) and was well isolated from other lion populations by surrounding deserts. Although often regarded as huge, the Barbary lion was no larger than most African lions and about the same size as East African lions (Mazák 1970). However, Barbary lions had darker and more grayish background coloration (resembling Cape lions in this regard), with long hairs (1.2–1.4 inches) in

Plate 1. The first team of Earthwatch volunteers (EW 2002-Team I) to participate in the "Lions of Tsavo" project, May 2002.

Plate 2. Collaborators in field studies of lions at Tsavo, from left: Roland Kays, New York State Museum, Albany, NY; Julie Thornton, Bradford University, U.K.; author; Samuel Kasiki, Tsavo Research Centre, Kenya Wildlife Service (photo courtesy of Chris Wiser, 2002).

Plate 3. The man-eater of Mfuwe, shot by Wayne Hosek, as it appears in the Field Museum diorama today. (Photo by J. S. Weinstein. Field Museum neg. #Z94328C. © The Field Museum.)

Plate 4. Adult male in Tsavo East National Park, with burrs matted in ruff and mane. Photo by J. S. Weinstein, October, 1998. (Field Museum negative # Z984029c, © The Field Museum.)

Plate 5. A maneless lion, "Bahati," being studied by Earthwatch volunteers on The "Lions of Tsavo" project. Photo by B. D. Patterson © 2003.

Plate 6. Kanderi pride female and two half-grown male cubs in Tsavo East National Park. Photo by J. S. Weinstein, October 1998. (Field Museum negative # Z985908c, © The Field Museum.)

Plate 7. Adult male in Tsavo East National Park. Photo by J. S. Weinstein, October 1998. (Field Museum negative # Z983932c, © The Field Museum.)

Plate 8. A scantily maned lion, "Cassius" (foreground), and part of his pride being studied by Earthwatch teams on The "Lions of Tsavo" project. Photo by B. D. Patterson © 2003.

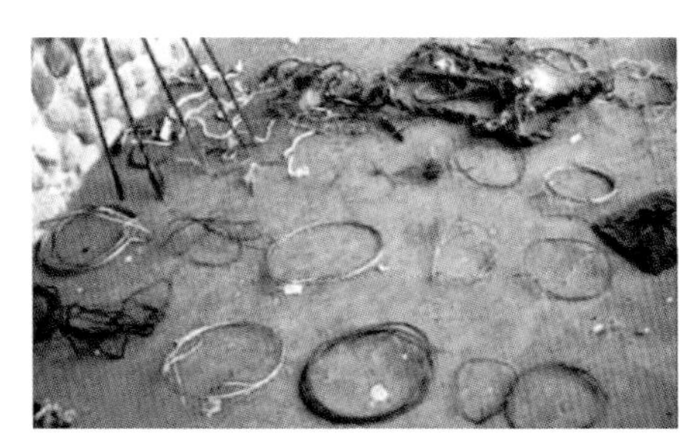

Plate 9. (Top) Young male from Masai Mara in snare set for bushmeat; (bottom) snares removed by Youth for Conservation volunteers in sweeps to eliminate poaching from national parks and reserves. (Photos courtesy of Josphat Ngonyo, 2002.)

Plate 10. Recording data from the male lion "Romeo" on Taita Ranch, Greater Tsavo Ecosystem. Individuals include Jeff Taylor (with camera), veterinarian Francis Gakuya, graduate student Alex Mwazo, author, and Taita Discovery Centre program manager Leigh Ekklestone. (Photo courtesy of Chris Wiser, 2002.)

Plate 11. The author fitting the male lion "Romeo" with a satellite-radio collar on Taita Ranch, Greater Tsavo Ecosystem. Photo by Jeff Taylor. (© B. D. Patterson 2002).

both sexes. The large and very long manes reached behind the shoulders well onto the back. On the abdomen, two parallel strips of long hair formed the belly mane, which thinned as it ran onto the chest. It differed from other lion populations both in social system and in breeding season.

Other African lions range over most of sub-Saharan Africa outside the humid rainforest belt. They occupy a diverse array of habitat types, from sand dunes in coastal desert to alpine moorlands. In the mountains of East Africa, they range up to 11,800 feet on Mount Elgon on the Kenya-Uganda border and reach 13,900 feet in the Bale Mountains of Ethiopia (Yalden, Largen & Kock 1980). They regularly surmount the most formidable barriers to distribution, routinely swimming across rivers as large as the Zambezi (Ansell 1960) and the Okavango Delta (Smithers 1971). Resident lions typically have enormous home ranges, but many lions are nomadic and apparently move over hundreds or thousands of miles in search of game and undefended territories. These wanderings carry them far beyond their normal range limits—lions are recorded deep in the Sahara desert (Smithers 1983). One ecological effect of such movements is that lions can recolonize even distant areas from which they had been extirpated (Anonymous 1998). An important genetic effect of this continual movement of individuals is to carry genetic variation from one region to the next. The evolutionary effect is to enrich the genetic diversity of local populations at the expense of regional differentiation.

Coupled with ecological and genetic characteristics that foster individual variation, lions also exhibit a widespread but fundamental developmental trend: Bigger mammals appear to be intrinsically more variable than their smaller counterparts. I first encountered this phenomenon in Canadian zoologist Stan van Zyll de Jong's superb review of the North American bison. The two living forms—plains and wood bison—differ remarkably in skull shape without appreciable genetic or physiological differentiation. This kind of variation can be confusing for even the most competent taxonomists. In the second edition of *The Mammals of North America*—which is *the* definitive reference for professional mammalogists—Professor E. Raymond Hall of the University of Kansas arranged the ninety-four names formally proposed for North American brown and grizzly bears into seventy-seven distinct species (1981)! Genetic analyses confirm that all not only are members of the same species but belong to the same species as Eurasian brown bears (Paetau et al. 1998). Their regional differentiation (which appears far more significant than that in lions) may reflect the more localized pattern of distribution and adaptation shown by bears. It may also be influenced by the impressive topographic variation that dissects much of their range.

Evolutionists Benedict Hallgrímsson and Virginia Maiorana recently documented this effect in an analysis of measurements from sixty-five thousand individuals belonging to 351 mammal species! Size-relative variability increases significantly with mean species body size, making larger species significantly more variable (Hallgrimsson & Maiorana 2000). The numerous scientific names for African lions reflect this same tendency for large mammals to exhibit more variation in shape—that variability makes them appear more highly differentiated, and hence deserving of a new name, unless one uses a relative "yardstick" of distinction.

Size Varies Geographically—and Individually

It is difficult to portray accurately the size of lions. True, individual animals can be measured and weighed with precision, but the meaning and significance of these measurements is unclear. First, lion sexes differ significantly in size, and each shows substantial local and individual variation. Second, from an evolutionary perspective, an individual is a temporary vessel, exposing a unique set of genes (its genotype) to a unique environment. It is populations and gene pools that evolve, not the individuals they comprise. In most sexual species, these individuals' sons and daughters will have different genetic constitutions and different physical attributes.

The ultimate physical or behavioral expression of any genotype (its phenotype) is shaped by its environment during development.* Different conditions can produce remarkable variation in the absence of genetic differentiation. For example, United States men raised in the mid-twentieth century are usually four to five inches taller than their fathers, and differences in childhood nutrition alone are sufficient to explain this difference. Cold War diets were far more nourishing than Great Depression ones, and sons realized more of the potential growth afforded by their genes.

Like people, lions are long-lived, with an extended maturation period and a lifestyle even more strongly affected by boom-bust "economics." So it seems reasonable to expect substantial developmental variation in lions. Several of the names proposed for geographic races of lions are based on exceptionally large or small individuals. Subsequent studies have confirmed that these blend imperceptibly into neighboring populations, making it impossible to apply those names to any objective entity.

As noted in Chapter 1, Tsavo lions are widely reputed to be larger than lions elsewhere in Africa—such claims fill popular books, magazines, and

*The great geneticist Theodosius Dobzhansky termed this envelope of developmental possibility the *norm of reaction* (1970).

travel brochures (Caputo 2000, 2002). However, this has never been scientifically corroborated—the assertion is apparently derived from the field measurements Patterson reported for his Tsavo lions at the turn of the century. He reported lengths and heights for the two man-eaters that would rank them among the very largest of lions. Adult male lions seldom exceed nine feet in length, and even the shorter of Patterson's lions rivals the longest of the 150 lions taped by Col. Stevenson-Hamilton in South Africa (Astley Maberly 1963). But Patterson gave similar size measurements for a lion he shot on the nearby Athi Plains southeast of Nairobi. Lions from this region are abundantly represented in museum collections and are large but unexceptional in size. Col. Patterson may have mismeasured his lions in some manner, but even if he did not, his own data demonstrate that lions from the Athi Plains and those from Tsavo are equivalent in size.

Table 5.2 contains measurements of body size for various populations of lions, placing the figures reported by Col. Patterson into a broader geographic context. Although they were undeniably large, the Tsavo man-eaters were adult males in good condition, and such lions are typically large. Still, a sample containing only four adult males from the Aberdares National Park in Kenya nearly exceeds the reported size of the Tsavo lions. Tsavo lions do

Table 5.2. Body size in lions (*Panthera leo*).

	Body + tail length (ft)	Weight (lb)	Source
	Males		
Kenya (largest lion shot)	9.74	412	Roosevelt & Heller 1914
Aberdares National Park, Kenya	8.2–9.5	396–506	Kenya Wildlife Service, personal communication*
Tsavo man-eaters	9.5–9.68	—	Patterson 1907
Selous Game Reserve, Tanzania	9.35	480	Rodgers 1974
South Africa	8.6–9.2	314–346	Smithers 1971
Gir Forest, India (*persicus*)	—	352–418	IUCN†
	Females		
Aberdares, Kenya	6.9–8.2	202–264	Kenya Wildlife Service, personal communication*
South Africa	8.1	234	Smithers 1971
Gir Forest, India (*persicus*)	—	242–264	IUCN†

*Sample based on adults among 22 animals culled from the Aberdares National Park by the Kenya Wildlife Service between July 1997 and March 1998.

† World Conservation Union, Species Survival Commision, Cat Specialist Group: http://lynx.uio.no/catfolk/sp-accts.htm for Asiatic lion

not rival the largest lions known from Africa, whose length Francis Harper gave as ten feet seven inches (1945). A South African lion is listed in the *Guinness Book of World Records* as having weighed 698 pounds!

Figure 9 depicts variation in skull size and shape in 176 male and female African lions at least four years of age; most represent localities in East Africa. Animals recently culled by KWS rangers in the greater Tsavo ecosystem are indicated by closed diamonds and the three Field Museum man-eating lions by closed squares. Neither male nor female lions from Tsavo appear exceptional in terms of skull size or shape. Each of the Tsavo lions fits comfortably within the envelope of variation defined by other African lions. In addition, the historic man-eaters are indistinguishable in size from those inhabiting the park today. As a size standard, the largest lion measured by Roosevelt and Heller on their 1909 safari had a greatest skull length of 16.0 inches and a breadth across the cheekbones of 10.7 inches (Roosevelt & Heller 1914).

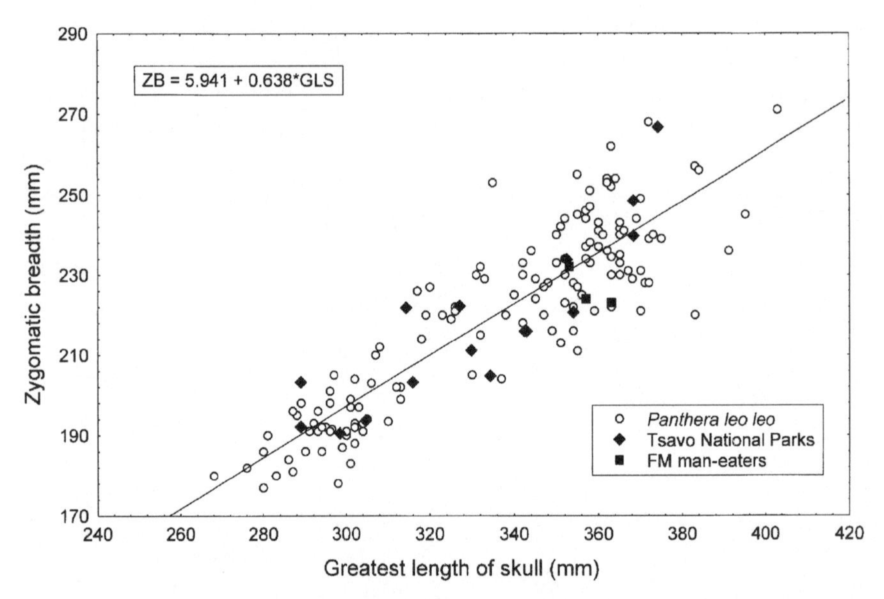

Figure 9. Scatter diagram of skull shape in some African lions. Plotted variables are skull breadth across the cheekbones plotted on skull length. East and South African lions show considerable variation in this character, lions plotting above the line having relatively broad skulls and those below being relatively narrow. A sample of lions from the vicinity of Tsavo East is included,showing that Tsavo lions do not differ in skull shape from other African lions, and that the variation present in Tsavo very nearly equals that evident elsewhere in East Africa (graph from Patterson et al., 1999).

In designating Tsavo lions as "buffalo lions," Gnoske and Kerbis Peterhans proposed that Tsavo lions have smaller heads but bigger bodies than lions elsewhere (Gnoske & Kerbis Peterhans 2000; von Buol 2000). To test this idea, body weights or measurements of skulls and of skeletons aside from the skull are needed from lions in Tsavo as well as in neighboring areas in East Africa; sadly, these data and resources are currently lacking. However, it is still possible that there are consistent differences in the proportions of Tsavo lions. During an April 2001 conversation with William Mukabane, a KWS ranger in Voi, he told me that he had preserved the skull of the largest lion he had ever encountered. As he dashed home to bring it, I reflected that Mukabane was an animal control officer who had shot more than 220 lions during his career—most in the greater Tsavo ecosystem (see Figure 10). This was clearly going to be an exceptional skull for my analysis. When he reappeared at the office with the skull, taken from an animal whose carcass had required ten men to load into a truck, I gasped. Its head measured only 13.5 inches long, about average for an adult male lion, including most of those from the trophy room in Voi (see closed diamonds in Fig. 9).

Maybe Tsavo lions never develop the massive heads of other East African lion populations, and their bodies become disproportionate in size. Maybe Mukabane was a just bad estimator of body size, or he and his crew were just tired the day they had to load that lion carcass. Surely, we need additional data and explicit analyses to decide this question.

Massive Reductions in Geographic Distribution

Until historic times, *P. leo* had one of the broadest distributions of all terrestrial mammals save humans and their dependents and hangers-on. Only twenty-five hundred years ago, lions ranged over most of Africa, except for the center of the Sahara Desert and the Congolese rainforests. Their range included coastal forests in North Africa, the Middle East and northern Greece, and southwest Asia well into the Indian subcontinent. But even that range was a vestige of its range at the end of the Pleistocene, when lions extended from the Cape of Good Hope to the North Sea and across all of Eurasia and the Bering land bridge east to Florida and south to Peru.

Even before the onset of human persecution, lions were never common in most senses of that word. This follows from the second law of thermodynamics, which states the tendency for useful energy to be lost as heat or entropy with each energetic exchange. Thus, green plants engaged in photosynthesis can capture only a portion of the sun's energy falling on them, and only a fraction of the energy in a plant's tissues can be assimilated by

Figure 10. The author, examining the skulls of three lions at the Tsavo Research Station that were shot by Kenya Wildlife Service rangers for attacking humans or livestock. Preservation of these samples by KWS enabled both morphological and genetic analysis of Tsavo lions. (Field Museum negative # Z984723c, © The Field Museum.)

an animal consumer. Apex predators like eagles and lions and sharks depend on numerous energetic transfers below them in the food web, each dissipating some of the energy entering the system. Traveling on foot before large-scale land conversion, Selous spent three years in the wilderness of East Africa before he even saw a lion (1881).

The end of the Pleistocene Epoch witnessed substantial climate change and massive turnover in prey, as many species of megafauna went extinct or suffered extensive range contractions. Shifting climates caused the replacement of the open steppes of Ice Age Eurasia by progressively more forested habitats. Increasing forest cover in Europe would have reduced its suitability for savanna-loving lions and sped their retreat into Asia Minor, the Middle East, and coastal Africa. Still, many authors, including Daniel Giraud Elliot and Francis Harper, maintained that humans began their systematic extirpation of lions at the close of the Pleistocene and were largely responsible for driving them out of Europe.

In northern Greece and Macedonia, lions were abundant in the fifth century BC, according to Herodotus. Lions ravaged the supply trains of Xerxes as he marched his armies across the Chalcidice Peninsula in 480 BC. Aristotle wrote that lions still ranged throughout that region a century later but were becoming rare. By the first century AD, most of the lions living in Mediterranean countries had been extirpated as Romans scoured the region for combatants for their games (Russell 1994). Over a period of only forty years, more than fifty thousand lions were captured and brought to Rome, many from North Africa (Fitzsimons 1919).

The disappearance of lions from the desert steppes of southwest Asia is harder to explain. Lions may have retreated from this region at the close of the Pleistocene when the region's megafauna disappeared—many of their prey species had been widely distributed in Eurasia during the Pleistocene but now occur only in Africa. Alternatively, lions may have been extirpated from apparently suitable habitat in Southwest Asia by conflicts with pastoralists. In any case, the earliest records indicate lions were widely distributed from Syria through most of Iran and Iraq to northern and central India (Pocock 1931). In India, they were found in large numbers in the states of Punjab, Haryana, Rajasthan, Uttar Pradesh, Madhya Pradesh, Gujarat, and western Bihar (Singh 1995). This range crossed much of India's longitudinal span and covered much of its northern half.

During the last two or three centuries, human population pressure inexorably displaced Asian lions from the open grasslands and woodlands that make up most of their range; the advent of firearms quickly extinguished any that remained. By 1888, lions had been eliminated from India except for the Gir Forest, where they took refuge in dense cover and land that was worthless for agriculture (Chellam 1996). This forest belonged to the Nawab of Junagadh and was used as his private hunting preserve. After the Gir lion population dwindled to a handful of individuals, the Nawab banned all hunting. Careful management led to recovery, so that in 1996 two hundred to three hundred lions lived in the 560-square-mile area (Chellam 1996; Jani & Malik 1997).

As discussed in Chapter 9, sport hunting has had a negligible effect on the conservation status of most animal species. By far the most critical threat is posed indirectly, by habitat destruction and conversion. For some, such as the American bison or the great whales, commercial exploitation has been decisive. Yet sport hunting has played a major role in the contraction of lion populations and range.* The American hunter Leslie Simpson is supposed to have killed 365 lions in a single year in East Africa (Hunter 1952), while a British officer shot 300 Asian lions during the 1857 Indian Mutiny (Pocock 1939). To place these numbers in context, only 20 Asian lions remained in 1900 (Chellam & Johnsingh 1993).† Local bounties and the fur trade added commercial incentive to the thrills of sport hunting.

But today in Africa, habitat destruction retains its customary position as the chief of Extinction's executioners. Burgeoning human populations require ever more space, and lions need space in which to live. In agricultural areas, lion populations are soon targeted for elimination. Their depredations on livestock and the threats they pose to humans quickly elicit retaliation, and farming preempts land needed by native prey. Herbivores that would otherwise feed on crops are removed or fenced out, leaving only smaller prey animals that are able to live in the interstices. This residue is often sufficient for solitary leopards but rarely suffices for groups of lions, accounting for the differential impact of humans on the two large felids (Bertram 1978).

Statistics are unnecessary to establish the quickening disappearance of lions throughout their remaining geographic range. One has only to examine the terminal dates of appearance, shown in Table 5.3. "There is probably no

*Some of the methods used were more sporting than others, and in one, hunting from horseback, the balance seemed rather to favor the lions. Although it facilitated the search for lions, reliance on horses sometimes placed hunters in a vulnerable position and caused several tragic incidents. One of the most famous involved George Grey, who hunted lions on the Kapiti Plains with Sir Alfred Pease, an accomplished lion hunter. When hunting on horseback, Pease never approached lions closer than two hundred yards, because a lion's sprint speed is much faster than a horse's, at least with rider and over broken ground. In the thrill of the chase, Grey rode to within ninety yards of a lion in flight, who then turned and charged, with fatal consequences (Johnson 1929). Even the most accomplished riders were vulnerable. A former officer in the Royal Hungarian Hussars, Fritz Schindelar was also dragged from the saddle and killed by a lion. But hunting lions is always risky, and about half of the professional hunters known to John Hunter had been attacked and mauled by lions (1952).

†This number is currently disputed. Although given by a knowledgeable gamekeeper, this estimate might have been artificially low to discourage further sport hunting (http://lynx.uio.no/catfolk/sp-accts.htm).

Table 5.3. Modern range collapse in the lion (*Panthera leo*), listing dates of last records in various countries within its modern range.

Country	Date of last recorded lion sighting	Source
Greece	350 BC	Guggisberg 1961
Azerbaijan	AD 900s	IUCN†
Israel	AD 1100	Guggisberg 1961
Samaria	AD 1200	IUCN
Libya	1745 (Pashalik) Tripoli	Guggisberg 1961, Harper 1945
Egypt	2nd half of 18th century	Guggisberg 1961
Pakistan	1810	IUCN
	1935 (Bolan Pass near Quetta)	Harper 1945
India	1814 (Bihar State)	IUCN
	1848 (Damoh District)	Harper 1945
	1856-58 (Delhi District)	
South Africa	1858 (Cape Province)	Harper 1945
	After 1898 (Johannesburg)	Guggisberg 1961
	Late 1800s (north of Orange River)	
	1865 (Natal)	
	1858 (Transkei)	
Turkey	1870 (near Birecik)	IUCN
Iraq	1870s (upper Euphrates)	Guggisberg 1961
	1910s (Mesopotamia)	
	1918 (on lower Tigris)	IUCN
Tunisia	1891 (near Babouch)	Harper 1945
Syria	1891 (near Aleppo)	IUCN
Algeria	1893 (near Batna)	IUCN
Niger	1918 (near Aïr)	Guggisberg 1961
Morocco	1922	Harper 1945
	1940s? (Atlas Mountains)	IUCN
Iran*	1942 (near Dezful)	Heaney 1943

*There are no confirmed records of lions living in Baluchistan or Afghanistan (Harper 1945)

† World Conservation Union, Species Survival Commision, Cat Specialist Group: http://lynx.uio.no/catfolk/sp-accts.htm for Asiatic lion

other species whose distributional range has shrunk over historic times to the extent shown by the lion" (Smithers 1983, 375). The process of converting wild lands into farms and grazing land for livestock continues throughout Africa, and this inexorably erodes and subdivides the remaining range of *P. leo.* Lions are difficult to survey, and estimates of their population sizes are nearly impossible to make. Were data available, several other African countries would certainly be added to the list in Table 5.3, and in dozens of others the threat to the continued survival of lions is real. The uncertain prospects of lions and other African wildlife are discussed further in Chapter 9.

Chapter 6
Hunting and Social Behavior

Lions are unique among cats in being social. Both males and females belong to long-lasting social groups, and these groups serve as the context for their hunting, territorial defense, mating, and cub-rearing behaviors. It is impossible to understand these features of lion biology without considering their sociality, and sociality has important tie-ins with the evolution of manes and the transmission of man-eating habits.

Hunting Behavior

Where prey is very abundant, as during Serengeti migrations, lions have little more to do than rouse themselves from a shady resting spot and amble toward the milling herds. In the best of times, any direction leads to paydirt. But where prey is scarce, lions may follow hunting circuits rather than use radial foraging trips from favored resting spots. Guggisberg believed that lions in Malawi followed a route that exceeded a hundred miles, permitting one to predict their movements up to a month in advance. In the Kalahari Desert, Eloff also determined that lions use circuits, although these are not fixed. One group of lions took four days to complete a circuit sixty-nine miles long (1973).

Lions hunt in groups, and groups of lions are more effective at capturing prey. George Schaller's classic study in the Serengeti showed that when lions hunted in groups, 30 percent of their stalks were successful, versus 17–19 percent kills when only a single individual was hunting (1972). In addition, groups can prey on larger animals than would be possible for lone hunters. Lions are formidable predators, but single lions cannot easily tackle African buffalo. The immense size, very thick hide, and low center of gravity make buffalo exceedingly difficult to tackle, even for several lions. Their deadly horns, herd-living habits, and willingness to dash back to the rescue of beleaguered fellows only heighten the difficulty of preying on buffalo. Buffalo herds do not panic, as do wildebeest or zebra, on sighting or scenting lions but instead take up a defensive formation bristling with piercing horns backed by massive sinews, gathering calves safely into the interior. Lions are never so reckless as to challenge these defensive formations (Rushby 1965).

But adult male buffalo often leave the herds, whether from disdain for the throngs or a predilection for choicer forage or more leisurely movements. Older bulls typically travel in small loose groups. Even when lions encounter a bull that is separated from his companions, a single lion is typically unable to overcome his defenses. But two or more lions may, as a lone buffalo can defend only its front end (Guthrie 1990). Kenyan outfitter Iain Allen witnessed two maneless male lions attacking a lone buffalo bull in Tsavo. The lions positioned themselves at opposite ends of the buffalo. As one engaged the front of the buffalo with a series of feints that kept it well clear of the buffalo's horns, the other savaged his hind end. When the buffalo whirled around to protect himself, the other lion pressed the attack. The buffalo finally backed into a dense acacia, making him all but impregnable, but not before sustaining a profusely bleeding wound. The lions retreated to shade and waited as the bull sat in the sun, bleeding. When his front end collapsed, the lions roused themselves and completed the kill. Killing buffaloes requires strategy and patience as well as strength.

Individual hunting behavior is flexible and varies with group composition and the behavior of one's associates. Nevertheless, lions appear to adopt distinct roles that they play in hunt after hunt. One of the most comprehensive studies was conducted in the open, semiarid plains of Etosha National Park in Namibia. P. E. Stander analyzed data from 486 coordinated group hunts to assess cooperation and individual variation in hunting tactics. Hunting groups generally consisted of lionesses that circled prey ("wings," or drivers) and those that waited for prey to be driven toward them ("centers," or ambushers). "Wings" frequently initiated an attack on prey, while "centers" moved only short distances and usually captured prey fleeing from other lionesses. An individual's adoption of these roles transcends any one hunt, and females seem to have a preferred role, based on the frequency with which they assume it. Hunts in which most lionesses occupied their preferred positions had higher probabilities of capturing prey (Stander 1992). Actions are coordinated by visual cues and low purring growls that seem to have a ventriloquistic quality.

Foraging success is not the only factor controlling the grouping patterns of female lions. Females also appear to band together to defend young against infanticidal males and to maintain territories over the long term (Stander 1992). Discussion of this subject will continue later in the chapter.

Lion Activity Cycles

Lions can be active by day or night. In any region, their activity schedule depends on the threat of human persecution and on their physiological tolerance. Even a century ago, it was clear that humans could alter the daily rhythms of lions, because Teddy Roosevelt noted, "If much molested they become strictly nocturnal" (Roosevelt & Heller 1914, 169). Persecution was sufficiently intense and widespread that the lion ceased diurnal activity over most of northern and western Africa during the twentieth century, but this effect was felt as far south and east as Zambia. Researchers in Garamba National Park, southern Congo, even cited the semidiurnal habits of the park's lions as a measure of the park's effectiveness as a sanctuary (Verschuren 1958). The cause of these shifts was obvious from observing the behavior of lions within game reserves. Inside park boundaries, lions hunted and lounged about in broad daylight, while neighboring groups outside those parks emerged from their refuges only after dark. With this shift in schedule, lions also become shyer and more secretive (Smithers 1971).

However, lions may also become nocturnal or chiefly so for physiological reasons. Desert areas characteristically experience pronounced daily temperature cycles. Daylight temperatures may exceed 100° F but fall to 60° F at night. This alternation creates great disparities in the heat and water costs associated with activity at different times. In an equatorial desert scrub like Tsavo, lions may become more exclusively nocturnal during the dry season. In southern deserts like the Kalahari, with scorching daytime temperatures, lions are strictly nocturnal during the austral summer, while on the fringes of the Sahara, lions may be exclusively nocturnal year-round (Eloff 1973; Guggisberg 1961).

Lions see extremely well in the dark, and their hearing and sense of smell are certainly as good as our own. But there is no consensus on the relative importance of sight, sound, and smell in their hunting, and none is apt to emerge. Laboratory trials could establish lions' neurological capabilities, but these would be mute concerning how that information factored into a lion's decision-making process. In the field, lions *strategize,* making it difficult to know when an object passes into their perception. Once, Carl Akeley was startled to step within three feet of a lioness before she moved (1920).

In many parts of their range, lions inhabit arid landscapes and water availability becomes an overriding ecological concern for predator and prey alike. Although the need for water constrains their behavior, schedule, and range, lions exploit water to their own advantage in hunting. In Tsavo, especially during the dry season, lions will lie in ambush near permanent water sources, knowing that prey are obliged periodically to obtain water there. But this strategy is effective only where and when water is critically limiting. Elsewhere, lions must be more resourceful to obtain their meals. As Roosevelt and Heller remarked, "But of the numerous kills we came across, several hundred in number all told, only a few were by the drinking-places. The great majority was out on the plains. Evidently the lion far more frequently kills his game by stalking, still-hunting, or driving on the plains than by lying in wait at a watering-place" (1914, 173).

What Lions Eat: A Variable and Adaptable Diet

To achieve its extraordinarily broad distribution, the lion has had to be variable and adaptable in its ecology and behavior. This variability is especially evident in its diet. The pantherine body plan permits a broad range of predatory habits, but these are augmented by the variable social system of lions. As a consequence, lions exploit a greater range of foods than is possible for their solitary relatives, like leopards. The sociality of lions also permits group transmission of accidental discoveries. If by chance a lion learns how to prey on an animal that is not ordinarily attacked, other lions can learn by example, leading to the establishment of a local tradition. This can lead to remarkable differences in diet in closely adjacent areas. Such learning is likely responsible for the differences between two prides in Botswana recently documented by zoologist Pieter Kat and photographer Chris Harvey. They determined that the third commonest item in the diet of the Mogogelo pride was baboon, whereas the adjacent Santawani pride ate no baboon at all. One of the Mogogelo lions had learned that treed baboons could be panicked into abandoning their safety in the treetops if the lion pretended to climb the tree, and this tactic was quickly adopted by other members of the group (Kat & Harvey 2000; Roosevelt & Heller 1914).

Lions are known to eat everything from lizards to elephants, even other lions.* Aardvark and hippo provide food for Congolese lions (Verschuren

*In *The Man-eaters of Tsavo*, Col. Patterson recounted a hunt in which two lions were left dead overnight where they lay—the next morning, two other lions were found feeding on one of these carcasses. Observers have also witnessed natural and uninterrupted instances of cannibalism—in one, a male killed a female in a squabble over a kill, then consumed much of her in preference to the prey (Fitzsimons 1919).

1958). Although lions are fearsome hunters, they eagerly appropriate meals wherever they find them. Hans Kruuk showed that lions in his study area more frequently scavenged from hyenas than the reverse—up to 80 percent of the carcasses lions were seen feeding from had been killed by hyenas (1972). Lions consume at least thirty-seven species of prey in South Africa's Kruger National Park, most falling between 100 and 600 pounds in body mass (Pienaar 1969). In most studies of lions, three or four prey species constitute most of the diet. For example, in the short-grass plains of Etosha National Park, sixteen species of prey are hunted, but 95 percent of the diet consisted of five species: springbok (621), zebra (135), wildebeest (56), oryx (16), and springhares (12) (Stander & Albon 1993). Lions living in expansive and productive grassland habitats seem to have more specialized diets. In a complex habitat mosaic like Botswana's Okavango region, lions prey on at least nineteen species, and eight species must be counted to reach 75 percent of their diet (Kat & Harvey 2000).

Some of this dietary variation is local, and doubtless tracks variation in prey abundance. Judith Rudnai found that lions in Nairobi National Park selected prey in approximately the same proportions as their relative abundance, suggesting that availability was crucial to prey choice (1974). In Kalahari Gemsbok Park, gemsbok is the lion's favorite meal, while zebra is the primary item in the menu of lions in Etosha National Park (Eloff 1973). For every species of prey except Grant's gazelles, at least 25 percent of the lion kills in the Serengeti were calves or juveniles. Profitability also varies with pride size: Larger prides have shorter handling times, and small groups appear unable to capture buffalo (1993). In the Kalahari park, where lions are highly dispersed and groups are small, Eloff found that small animals and juvenile antelope make up 50 percent of the diet. Porcupines ranked high in the diet (the lions have a stylized manner of dismembering and consuming them) (Eloff 1964, 1973).

"Buffalo lions"

Recently, zoologists Tom Gnoske and Julian Kerbis Peterhans proposed that Tsavo lions are "buffalo lions," specially adapted for and especially dependent on herds of African buffalo, *Syncerus caffer* (von Buol 2000). Although their technical presentation of this idea and its supporting data are still forthcoming, the possibility they raise is an intriguing one. Both park rangers and tour operators in Tsavo invariably note the association of lion prides and buffalo herds. During our survey and inventory of Tsavo lions in 1999, Roland Kays and I habitually studied the spoor of buffalo for any sign of lions that might be trailing them. Certainly, one frequently finds groups of lions lounging near the carcasses of buffalo, but such observations offer little to overall understanding of their diets. Lions might just as frequently eat small, inconspicuous prey like hares and dik-diks and depend on them as key food resources, but it is far harder to observe and document such reliance.

To date, little has been published concerning the diets of Tsavo lions, and not much more is known of their behavior and social system. In the only quantitative study of diet yet published, Kenyan biologist Ayeni documented carcasses of animals killed near waterholes in Tsavo. He found nineteen buffalo, fourteen eland, eleven waterbuck, seven kudu, six zebra, four oryx, two kongoni, and one giraffe, and attributed the majority of these kills to lions. This frequency of prey species in his samples did not appear to be correlated with their relative abundance in Tsavo. Rather, Ayeni believed that lions prey preferentially on ungulates that visit waterholes at the same time of day that they do (1975).

On the Kapiti Plains, 120 miles northwest of Tsavo, lions were commonly observed on the carcasses of kongoni and zebra but were rarely associated with buffalo. Roosevelt and Heller believed that lions preyed easily on zebras and hartebeests but that buffalo were too formidable for them to challenge. Isolated observations like this one support the notion that some lions—perhaps including those in Tsavo—may specialize on buffalo, while others eat mainly wildebeest and zebra. But quantitative data on lion diets from all corners of Africa show that most lions feed on buffalo, some of them preferentially so. Ironically, given the supposed distinction between "buffalo lions" and "pride lions," David Scheel found that it usually takes a pride of lions to bring down a buffalo, because small groups are seldom willing or able to tackle them (1992).

For example, in the Selous Game Reserve in southeastern Tanzania, Rodgers determined that lions killed wildebeest (34%), warthog (25%) and buffalo (22%) in declining numbers. However, preference ratings, which assess prey selection in relation to abundance, show that buffalo are most highly preferred—3.4 versus 3.2 for warthog and 1.3 for wildebeest (1974). In Kafue National Park, Zambia, buffalo were both the principal and the most frequent prey of lions (Wilson 1975). Frank Ansell gave the diet of the lion in Zambia as "medium to large mammals, especially buffalo (*Syncerus caffer*), of which fully adult males are often preyed upon" (1960, 48). Buffalo were the most frequently represented of twenty-one species making up lion diets across all of Hwange National Park in Zimbabwe (Wilson 1975). In Chobe National Park, Botswana, buffalo, zebra, and hartebeest make up 50 percent of the diet year-round, and buffalo are focal species in both rainy and dry seasons (Viljoen 1993). And in the savanna woodlands in South Africa's Kruger National Park, the main prey species of groups of male lions, especially for nonterritorial males, was buffalo. Female lions in Kruger preyed more frequently on the most abundant medium-sized ungulates, wildebeest and zebra (Funston et al. 1998)

Do all of these studied populations represent this cryptic "buffalo lion"? No, because even the "pride lions" of the Serengeti—the counterpoint to buffalo lions—feed heavily on buffalo. Detailed studies by David Scheel and his associates show that buffalo are principal prey of prides during the dry season, when migratory herds have departed. Buffalo are also important to lion diets throughout the year in the woodlands, where seasonal changes in prey abundance are damped. There only larger prides attack buffalo, because of the difficulty and danger that is involved. To lessen the risk of counterattack and injury or even death, lions preferentially attack solitary males rather than herds (Packer, Scheel & Pusey 1990). Studies in nearby Lake Manyara National Park show that 88 percent of buffalo are killed by lions, and male buffalo have a predation risk three times higher than that of females, presumably because they are less closely and frequently associated with herds (Prins & Iason 1989).

Lions then must be seen as extraordinary opportunists that exploit a wide variety of available prey. Their exploitation patterns reflect seasonal abundance and migration patterns of prey, habitat structure and its effects on prey grouping and spacing, and lion group size and social context. As Alan Turner observed, there is great flexibility in the hunting behavior of large predators generally, so that neither the elephant-hunting lions of Botswana's Chobe National Park nor the buffalo-hunting lions of Tanzania's Lake Manyara National Park offer a typical picture (Turner & Antón, 1997). The common denominator of these strategies is the potential of lions to adapt their behavior, social system, and diet to whatever circumstances they encounter. Throughout their geographic range, opportunism in feeding brings lions into conflict with people, either causing serious losses in cattle-raising areas, for example (Ansell 1960), or attacking people themselves. Such conflicts are multiplied and intensified by the continued fragmentation of lion ranges. (These issues are discussed in more detail in Chapter 9.)

Social Biology: The Pride

The fundamental unit of lion society is the pride. *Pride* refers to a permanent social group of two to eighteen adult female lions and their dependent offspring. Prides may occupy the same areas for generations. In the early

1980s, Anne Pusey and Craig Packer found that the Masai pride in Serengeti National Park, Tanzania, still patrolled the same area it had occupied when Schaller initiated studies there two decades earlier, although no original pride member still survived (Pusey & Packer 1983). Prides patrol and defend their territories against incursions by other prides, as well as by nonterritorial lions. Pride territories are perhaps best characterized as the areas exploited and defended by hunting groups.

Studies in the Serengeti show that virtually all of the females in a pride are born to it. Most of their surviving female offspring remain, but roughly a third leave. Young females often depart in groups, generally when new males join the pride or when females give birth. They may also leave singly to become nomadic, but dispersing females rarely join other established prides (Pusey & Packer 1983). Female groups are closed societies, and those living beyond the pride's limits face delayed reproduction and/or heightened mortality (Pusey & Packer 1987).

Female lions can conceive at as early as 2.5 years of age, and there is generally a 2-year interval between litters. Lionesses sometimes abandon litters consisting of a single cub, which enables them to reenter estrus and achieve higher reproductive output (Rudnai 1973). In Nairobi National Park, Judith Rudnai documented matings throughout the year, but births peaked in December and January. Gestation lasts about 110 days. Immediately prior to birth, females leave the pride and seek a secluded place to give birth. Caves, abandoned warthog or aardvark burrows, or simply tall grass or brush may be used for shelter. The average number of young per litter is three, with a balanced sex ratio. The female and her cubs do not rejoin the pride for 4–8 weeks, and then only if there are no older cubs already in the pride (Plate 6). Coordinated timing of births is crucial, because females suckle any cubs within the pride, which leaves smaller cubs very much at a disadvantage. Cubs suckle for 6–7 months (Rudnai 1973). Giving birth alone and repelling male intruders are highly risky behaviors for solitary mothers, but both may be necessary to prevent infanticide.

Details of the social system of lions reveal their solitary ancestry. Unlike more highly social wolves and mongooses, lions have no dominance hierarchy. Nor do females suppress the reproductive cycles of other females in the group. Various features of lion biology invalidate the strategies that other species use to commandeer the reproductive options of their associates. As a result, female relationships are remarkably symmetrical and pride companions produce similar numbers of surviving offspring. Overt fights between females are simply too costly, as they could lead to crippling injury to oneself or a valuable companion. Females leave the pride to give birth and do not rejoin it until their cubs are less vulnerable, preventing them from being killed by manipulative females. Female lions contribute to communal care only when they have cubs of their own. And as Packer and colleagues have observed, this places a premium on reproductive synchrony and parity among females, because any attempt to eliminate companions as mothers inevitably removes them from being aunts and wet-nurses as well (Packer, Pusey & Eberly 2001). Female companions are hunting partners, co-defenders of territories, and nursemaids, not to mention frequently being kin.

One or more males are usually associated with these permanent groupings of females. Like the females, the males are often related—often either brothers or cousins—and live in lifelong associations. However, the composition of male groups is not so easily categorized. All of the males born into a pride eventually disperse from it. Typically, they are driven out en masse by aggression from older pride males during their second or third year* and rarely show sexual interest prior to emigration. Thus, dispersal is sex-biased: Females are sedentary and males disperse, which is characteristic of mammals and serves to prevent inbreeding.† Because males must head off to seek their reproductive fortunes elsewhere, there is little risk of incest and inbreeding depression in lions.

Thus, an emigrating cohort of males consists of brothers and cousins that leave the natal territory together and may remain so for life. But this period is one of heightened mortality, as the dispersers must learn to hunt for themselves and avoid potentially lethal encounters with territory holders of both sexes, who control the best hunting grounds in the region. Seasonal resource bottlenecks, such as disappearance of grazing herds from the Serengeti during the dry season, with the concentration of predators into the adjacent woodlands, take an especially severe toll on these nomadic groups. Half to two-thirds of juveniles fail to reach maturity, and other lions are the principal reason—through either infanticide or territorial clashes (Rudnai 1973). Heavy juvenile mortality creates occasions in which unrelated males may take up an association together, producing amalgamations of orphaned youngsters, survivors of pride takeovers, and/or nomads.

*Fathers may tolerate the presence of their three- and four-year-old sons, but this is rare among unrelated individuals (Pusey & Packer 1983).

†Curiously, the reverse situation (sedentary males and peripatetic females) is characteristic of birds. Both male mammals and female birds are heterogametic, meaning they have XY chromosomal complements. The implications of this finding for genetic structure and sexual selection were considered by Atmar (1991).

Sometimes young males will join much older ones, already past their prime, who are also relegated to life on the margins.

The size of lion prides is variable. Generally, prides are larger in the plains and where game is plentiful. In the Serengeti, the number of adult females in prides documented by George Schaller in his classic study ranged from 2 to 11, with a mean of 5.9. If subadult females are also tallied, the range is from 2 to 13 and the average 6.5 females (1972). In Kruger, Smuts gave a mean of 11.8 (1976). Pride size and hunting ability are key factors influencing hunting success. Only larger prides can tackle the largest and most dangerous prey (Scheel 1992).

Prides tend to be smaller in dense bush and where game is scarce. "In the Sudan, ... prides are mostly met with in twos and threes. Prides of more than 4 or 5 are definitely rare.... The same applies to Chad and the Central African Republic" (Guggisberg 1975, 151). In Botswana's Kalahari Gemsbok Park and Namibia's Etosha National Park, Eloff recorded an average group size of 4.2 individuals (1973). In Chobe National Park, on the humid eastern edge of Botswana, Child recorded one pride of 6, but typically only single or paired individuals (1971). The average size of fifty-eight lion groups that Wilson observed in Hwange National Park, Zimbabwe, was 3.8 individuals (1975), while in Tanzania's Selous Game Reserve, Rodgers found that lions live in groups having a mean size of 3.4 animals, based on 298 individual sightings (1974).

In Indian lions, most prides contain only two females, but some contain as many as five. Coalitions of 1-3 males defend territories containing one or more groups of females but associate with them only during mating and when scavenging from a large kill (Singh 1995). Male coalitions patrol territories two or three times larger than those of lone males. Small group size in Indian lions may relate to the small size of their typical prey, as chital deer weigh only 110 pounds. It is anomalous in terms of the high density of lions in the Gir—at one lion per 2.7 square miles, density there is equivalent to the upper range of estimates for lions in sub-Saharan Africa.* However, the Gir Forest lacks large numbers of predatory hyenas, and protection from infanticidal or carcass-thieving hyenas is one of the selective advantages favoring large group size in African lions.

Seasonal variation in pride size is known but poorly documented. Astley Maberly stated that prides in southern Africa numbered from 4 to 20 individuals, but aggregations of up to 30 individuals could occasionally be found. These larger groupings were usually noted during times of

*http://lynx.uio.no/catfolk/sp-accts.htm for Asiatic lion

resource scarcity and contained individuals of all ages and both sexes, although they were typically attended by only one prime male (1963). According to John Hunter, lions in Tsavo's rainy season leave their usual ranges to travel great distances, often singly or in pairs rather than in prides. During the dry season, when water sources again become limited, these smaller groups apparently coalesce back into more typical prides. Similar observations have been made by Tsavo's tour operators on lion groups seen near major water reservoirs like Aruba; aggregations of between 20 and 30 lions have been noted by Aruba Dam during the dry season. Hunter also noted that there may be "several males in a pride, each with his own harem, but there is generally one head male and the others defer to him" (1952, 25). Neither fission nor fusion of lion groups nor the existence of paramount males has been the subject of prior scientific study, though both may characterize Tsavo lions.

Male Coalitions and Mating Habits

Studies in the Serengeti indicate that male coalitions are an integral feature of lion society. Of the fourteen prides whose size he accurately determined, George Schaller stated "Each pride consists of 2 to 4 adult males, several adult females, and a number of subadults and cubs" (1972, 35). Females mate preferentially with resident males, and only cubs of females living in prides have much chance of surviving. Consequently, it is essential that males become resident, and this almost invariably involves contests with incumbents (Pusey & Packer 1983). Larger coalitions are more likely to establish residency nearby, and they acquire prides of females at younger ages. However, the largest coalitions are unsuccessful, because groups larger than five males usually split shortly after emigrating (Pusey & Packer 1987).

The average tenure of a male coalition with a pride of females is only two years, but the variation is considerable, lasting from a few months to as long as six years (Bygott, Bertram & Hanby 1979). Brian Bertram was the first to recognize the connection between this short-term association and the seemingly aberrant behavior of infanticide. Male lions have long been known to kill suckling cubs sired by prior coalitions, and the mothers of these cubs quickly come back into estrus, as lions are aseasonal breeders and induced ovulators. Bertram postulated that cub killing constituted a male reproductive strategy driven by the short tenure of pride males. This interpretation has been confirmed in subsequent studies on Serengeti lions. Males rarely reside with daughters that have grown to reproductive maturity, typically abandoning such prides and taking over others. The likelihood that a male coalition will abandon one pride for another depends on the age of cubs—younger cubs are more vulnerable to infanticide and require more paternal protection (Pusey & Packer 1987).

But infanticide may not be a universal feature of lion prides. Pieter Kat observed females in two prides living in Botswana's Moremi Game Reserve being mated by two or more male coalitions. Moremi males also seemed more tolerant of other males in their territories than is typical of Serengeti lions. With uncertain paternity, a male practicing infanticide might very well be destroying his own descendants rather than those of a rival (Kat & Harvey 2000). Infanticide is even rarer in other cats, where gestation is only 60 percent as long and interbirth intervals may be a quarter what they are in lions (Natoli 1990). Infanticide can prevail only where paternity can be assured (through territorial exclusion) and where the costs of cuckoldry are unsupportably high (because tenure is brief).

Males in a coalition seldom fight for copulating rights with females. Mating couples usually seek seclusion, but sometimes these trysts attract other males. In the case of coalition-mates, two males will often share a female without aggression. Many coalitions contain males at equal stages and conditions of development; this and their evident affection for one another may mitigate potential conflicts over access to mating (Pusey & Packer 1983). Commonly, the interloper takes his turn with the female during rests by the mated male, who continues his attentions unabated. At other times, he may be waiting to rejoin his companion, not to mate (Schaller 1972). Throughout the mating period, lions go without food and usually will not hunt (Eloff 1973). In a sample of some two hundred matings witnessed, intervals between copulations ranged from 4 to 148 minutes, averaging 17 minutes (Rudnai 1973). By providing for alternative parentage for the cubs, coalitions enhance the uncertainty of paternity that most animals face.

Why should lions mate hundreds of times during the three- to four-day period in a female's estrus cycle when she is receptive?* This is only partially answered by the fact that only one in three copulations results in fertilization. By lowering the probability that any copulation will result in conception, females may enhance the odds of encountering a highly fit male. Sperm competition mediates this "talent search," deciding which of the millions of ejaculated sperm will achieve fertilization. It ensures that only the healthiest, most active sperm reach the eggs. Promiscuity may also

*Estrus lasts from four to sixteen days, and the interval between estrus periods varies from a few days to over a year. There is a post-partum estrus, but conception happens only if the earlier litter is lost, resulting in the birth of another litter within four months (Smithers 1983).

help to limit infanticide, encouraging males to desist from killing cubs that might conceivably be their own (Hunter 1999).

Besides providing females with high-quality sires and protection for their offspring, males contribute importantly to a pride's hunting success. We have noted that prides hunting with males are often more successful in subduing larger, more valuable prey. They can also defend carcasses better against aggressive competitors like clans of hyenas or neighboring lion prides (Schaller 1972). Studies by Funston and colleagues in South Africa show that the proportion of time territorial males spent with and thus scavenged from their pride females was influenced by vegetation structure. Males spend more time with pride females in open habitats. Vegetation structure may change food availability and hunting success, territory maintenance costs, and/or defense of cubs (1998).

Male and Female Territories

Although male and female lions are both social and territorial, their territories differ in size, location, and purpose. For a female, territory is an inherited birthright, a place for securing food and rearing young. The unparalleled sequence of studies on Serengeti lions has shown that prides retain a recognizable territory even after all the original territory owners have died. For a male, the territory is a region containing females with which he can potentially breed. Territories of males can encompass the hunting ranges of various prides of females, and prides of females may hunt over the territories of two or more male territories. Besides such resident groupings, there are always nonresident lions, or "nomads," that drift through the landscape, concealed by the vastness of a lion's territory (Guggisberg 1961; Kat & Harvey 2000; Schaller 1972).

Territory size varies with both lion and prey density. Territories range from being very large in sparsely populated areas like Etosha, where densities are as low as 0.01 to 1.8 lions per hundred square miles (Stander 1990, 1991). Territories are smaller in more densely settled regions such as Ngorongoro Crater, where there are about 100 lions in one hundred square miles and territory size averages about thirty square miles (Elliott & Cowan 1978). During the lean season in Uganda, range size of lions is negatively correlated with prey abundance (Van Orsdol, Hanby & Bygott 1985). However, changes in territory size with food density are not automatic. Bertram found that territories in Serengeti woodlands don't change seasonally in size or location, despite doubling or tripling of the prey biomass they contain when migrating herds return (1973).

Territorial Markings

Spatial overlap between lion groups is almost inevitable, given the enormous areas they exploit. Individuals have been known to travel up to thirty miles per night, placing a premium on communication with members of their own and other groups. Lions use various signals to communicate their presence, maintaining contact with group-mates and avoiding potentially deadly confrontations with other lions. Signaling enables multiple groups to utilize the same piece of territory, albeit not simultaneously. Most of these advertising signals also contribute to territorial maintenance, especially scent-marking and roaring. Lions reveal their presence by uttering low grunts or soft moans during movements, periodic roaring, making tracks, scent marking and defecation, and frequent urination. By advertising the presence of lions, all these signals serve to alert others of potentially damaging confrontations that might arise from multiple use of a given area. Lions patrol their territories during hunting and searching for mates; they eject intruders as the situation warrants. While trespassers may receive a brutal reception, they may also be tolerated and ignored (Bertram 1973; Eloff 1973).

Lions leave scent by tracks, urine, and feces left in their activity areas. Anyone who has smelled cat urine knows that it contains plenty of volatile components that have the potential for copious biological information. Kristian Anderson and T. Vulpius used a gas chromatograph-mass spectrophotometer to characterize these components. The lion urine samples they analyzed contained fifty-five compounds, seven with the potential for species identification (1999). Lions use urine and scent from anal glands beneath the tail to mark various features within their territories. Spraying vegetation is mainly a male behavior and usually performed by mature animals. On the other hand, scraping—raking the ground, usually punctuated with urination—is common to both sexes once they are at least two years old. Feces are so casually deposited within the territory that Schaller believed they could not serve a purpose in communication (1972).

Roaring

Roaring is indisputably important to lion spacing and territoriality and has also been well studied. Both sexes roar, and they often do so in unison. Presumably, louder, more vigorous groups can defend a larger territory (Bertram 1973). Male roars serve as advertisements that the territory is defended, as males issue roars in territorial challenges. Shortly after we radio-collared the adult male lion "Romeo" for our Earthwatch project, on Taita Ranch in April 2002, he became very shy, being unhabituated to either vehicles or observers. Broadcasts of male roars were the most effective means of establishing contact with him, to lure him from the dense *Acacia-Commiphora* woodlots of the ranch. Broadcasting a typical roar sequence from our vehicle, we found that Romeo would be roaring synchronously with our tape by the third pulse and commonly moved quickly and with purpose directly toward us. (Our challenge, of course, was maintaining the illusion that it was another lion, and not our Land Rover, emitting the sound.)

In the most extensive studies of this response, behaviorists Jon Grinnell and Karen McComb found that when playback experiments are used to simulate intruder lions to pride males, it elicits cooperative behavior from the coalition. Whenever two or more resident males were present, they approached the apparent intruders. Residents failed to respond only when a single male faced the roars of three intruders. Cooperation in approaching the roars of strange males does not depend on kinship or on the behavior of one's companions but instead is based on mutualism (Grinnell, Packer & Pusey 1995). Numerical assessment skills likely evolved in response to avoiding the costs of fighting with larger groups (McComb, Packer & Pusey 1994).

Roaring serves additional group-specific functions. Female lions roar to maintain contact with their pride-mates and to defend their territory against other female groups. Playback experiments show that females can distinguish pride-mates from strangers based solely on vocalizations. They can also assess the odds of winning a territorial contest by comparing the number of intruders with the number of defenders. Serengeti females approach intruders aggressively only if they outnumber them.* In the Ngorongoro Crater, lions live at much higher densities, because prey species there do not migrate. Playbacks showed that Crater lionesses show greater aggression, approaching "intruders" more quickly even when the odds of winning are low. It may be impossible to avoid intruders or defer territorial defense when population density is high (Heinsohn, Packer & Pusey 1996).

But roars reveal the locations of females to potentially infanticidal males and here may serve an intersexual purpose. Playback experiments show that nonresident males are in fact attracted to female roars; they will approach roaring females but not males. Alien males approach a lone roaring female

*Adult lionesses display an unexpected variety of behavioral strategies when defending the group territory. Most cooperate unconditionally, but others contribute only when needed, and some "cheat" by consistently lagging behind and avoiding risks. These personality traits emerge in juvenile lions and remain consistent throughout their lives (Heinsohn, Packer & Pusey 1996).

more readily than they do a chorus of three, and single males are more hesitant to approach than pairs (Grinnell & McComb 1996). Females with cubs can sometimes successfully repel new male groups and are more effective in doing so than lone females (Pusey & Packer 1983). Consequently, nonresident males must be cognizant and wary of both males and females they encounter. By roaring in a chorus, female groups may minimize such encounters.

Other Communication Methods

Australian ecologist Rob Heinsohn and colleagues studied the development of group territoriality behavior in lions. They showed that juvenile females became progressively more likely to join adult females in challenging simulated intruders. Juvenile males showed no such tendency. As with adults, the willingness of females to rally depended on the number of defending adults as well as on the number of intruders. Differences between the sexes reflect the ultimate value of the territory to each group: Females stand to inherit the territory and their survivorship and reproductive value depend on it, whereas juvenile males will inevitably leave it within a year or two (Heinsohn, Packer & Pusey 1996).

Lions also use a wide range of facial expressions and body postures to communicate with one another visually. Some of these are important in interactions within the pride or coalition, others in territorial situations. A few serve to expose the mane, the subject of Chapter 7. Head rubbing and social licking are forms of tactile communication among lions. Both are frequently expressed and seem to promote social cohesion. Finally, cats produce a wide array of vocalizations, from purring and growling through roaring* (Schaller 1972).

Cooperation Among Lions

Cooperation is rare in nature. In the race to leave as many descendants as possible, individuals more commonly find themselves in direct competition with each other. Cooperation arises in situations where individuals can expect some return on their helpful behavior. In fact, it is the rule among close kin. However, as evolutionary theorist Sarah Legge noted, cooperation

*Schaller divided these into puffing, purring, bleating, humming, roaring, woofing, grunting and soft roaring, loud roaring, meowing, and growling-coughing-snarling-and-hissing, and identified the social contexts in which they are uttered (1972).

is also expressed among unrelated individuals in three contexts: group selection, reciprocity, and mutualism (1996).

Lion biology is full of opportunities for cooperation. Female lions hunt together, raise and suckle cubs together, jointly defend them from infanticidal males and other predators, and establish, patrol, and defend a territory together. Male lions hunt together, acquire a pride together, defend it from other males, and acquire, patrol, and defend their territories—some even acquire mates together. None of these activities takes place in an ecological vacuum. Consequently, a split-second decision about whether or not to rally to territorial defense may affect cub survivorship, immediate hunting success, and prospects for surviving the next season of food scarcity. Is it any wonder that scientists remain undecided over what factor caused lions to become social?

The classic explanation for sociality in lions is cooperative hunting, based on the heightened success and broader range of potential prey that group hunting allows. Group hunts *are* more frequently successful, and groups can certainly target larger prey. In some cases, success in hunting the principal prey species also increased linearly with lioness group size (e.g., Stander & Albon 1993). In these instances, the most important variables affecting the outcome of a hunt are group hunting and coordinated stalking. By determining per capita food intake in this way, cooperative hunting can result from selfish genetics.

But other evidence suggests that lion group size is not always optimized based on hunting. In a broad review of cooperative hunting, Craig Packer and colleagues concluded that cooperative hunting could as easily be a consequence of sociality as its cause. They found that cooperative hunting was favored when individuals faced a low probability of securing large prey on their own or when groups tackled small prey (Packer & Ruttan 1988). On the other hand, "cheating" was favored when individual hunting success is high and group size is large. Studying the predation patterns of Serengeti lions, David Scheel showed that they selectively cheat their pride-mates in hunts. Males cheat more than females, and lions cheat their pride-mates more frequently in chases involving prey that are easy to catch, where their failure to participate or help is less apt to be decisive (1992).

Extensive observation of female lions, again in the Serengeti ecosystem, show that lions' group size is not optimized for hunting success. Foraging success during the season of prey scarcity is highest in groups containing a lone female or five to six females. Yet despite lowered hunting success, radio-collared females sought to forage in as large a group as possible, contrary to the prediction that they should forage alone. Because mothers keep their cubs in a crèche and deploy maternity groups to defend them against infanticidal males or hyenas, a larger group is adaptive. Moreover, larger groups enable females to prevail in territorial disputes with neighboring prides and interlopers (Packer, Scheel & Pusey 1990).

Despite continuing debates, neither hunting, group territoriality, nor cub defense could ever be the exclusive cause of sociality. The importance of these factors shows remarkable variation throughout the lion's extensive range. They also vary temporally in any one location. Therefore, demonstrations that lions behave this way or that need to be qualified by the nature of the sample being studied.

In *Prides: The Lions of Moremi,* Pieter Kat and Chris Harvey also emphasized the extraordinary adaptability and variability of lions. Scientists and the lay public have come to know and understand lions far better than most other mammals, owing mainly to an exceptional series of studies in the Serengeti ecosystem initiated by George Schaller in the 1960s and continuing today. These studies have given us a wealth of information on such elusive biological properties as lifetime reproductive success and metapopulation demography. Yet Kat and Harvey contend this work has created an orthodoxy around lion biology that applies poorly to that species elsewhere. The challenge now lies with lion investigations elsewhere to produce data of comparable quality and meaning. In South Africa, Funston and colleagues recently showed the signal importance of hunting by male coalitions (Funston, Mills & Biggs 2001). And Kat has recently documented pride relationships in Okavango that are far less rigid than those in the Serengeti (Kat 2000; Kat & Harvey 2000).

Roland Kays of the New York State Museum, Samuel Kasiki of the Kenya Wildlife Service, and I recently initiated studies to document lion ecology in the greater Tsavo ecosystem. (Some of this work is described in Chapter 9.) Clearly, Tsavo represents an extreme environment for lions—it is hot throughout the year and exceptionally dry over an extended rainless season. Can we simply predict how Tsavo lions behave? For example, low prey densities in an unproductive environment like Tsavo's should result in low lion densities and small group sizes. (Karl Van Orsdol and colleagues substantiated that relationship in their Ugandan study [Van Orsdol, Hanby & Bygott 1985].) And during Tsavo's dry season, we might expect lion home range size to double, based on Viljoen's findings in Chobe National Park. However, our initial field studies show that both these predictions are false, and the answer can only be that the responses of these populations are individually tailored to their environments. This example shows that we've only begun to understand the variability and adaptability of this cat.

Chapter 7
The Lion's Mane: Geographic and Individual Variation

Adult males have a mane of long hair, up to about 16 cm [6.3 inches] in length, on the sides of the face and top of the head which extends on to the shoulders, around the neck and for a short distance down the spine. In subadults, this mane is always sandy, yellowish or tawny, but in some, with advancing age, it becomes black.
(Smithers 1983, *The Mammals of the Southern African Subregion*, 374)

In the book *The Carnivores*, Rosalie F. Ewer called the lion's mane "the only really striking example of sexual dimorphism in the Carnivora" (1973, 78). The mane symbolizes power and authority, both to humans and presumably to lions. For centuries, human rulers have adopted manelike mantles as trappings of power. Maned lions are practically ubiquitous in European heraldry, which is remarkable in view of the fact that lions have not lived in Europe for ten millennia. Guggisberg traced this symbolic appropriation back to the Crusades, when Europeans became aware of Middle Eastern lions—Henry the Lion, Duke of Saxony, included the lion on his coat of arms in 1144 (1975). And Jonathan Kingdon has noted the sexism that accompanies depictions of lions: "In fact the lion crops up in the symbolism of a great many civilizations. It is almost as universally present as the lioness is absent and it has to be admitted that the mane seems to alter human perceptions of this big yellow cat fundamentally" (1977, 368).*

*This symbolic role of the lion in Western civilization is especially interesting in view of its minor place in African art and iconography. "Of the animals represented in African art, the leopard probably appears more frequently and widely than any other, with the elephant as its only possible competitor. The lion, largest of cats, is strangely out of the running" (Roberts 1986, 97). In some places where lions do appear, as in the royal arts of the Fon and Akan peoples, they may actually be borrowed from or reinforced by European heraldry.

Because lions are unique among terrestrial predators in having a pride-based social system, manes are commonly attributed to selection pressures that arise in this unusual social milieu. In lions, the enduring social unit is a group of females—daughters, sisters, cousins, and aunts—together with a smaller number of adult males. Group membership of males—both those born into the group and those reproductively attached to it—is transitory. In this context, the mane is thought to function in threat and display, protect vital areas in fighting, and perhaps assist in individual recognition. The impacts of both physical and social environments on the development of this trait offer insights into lion biology and ecology.

Natural vs. Sexual Selection

Lion manes carry all the hallmarks of a sexually selected trait. A bit of background on the effect of selection on natural populations will help to clarify this. Selection on natural populations falls into two main categories: "natural selection" and "sexual selection."

Natural selection was the central organizing principle in Charles Darwin's *Origin of Species*—the agency that sculpts random variations of form or behavior into exquisite adaptations that fit an organism to its environment. Natural selection operates by granting individuals differential access to and success in reproduction. Traits and behaviors that favor the growth, survival, and success of individual *genotypes* (the suite of genes of each individual) are adaptive and proliferate in the population. Fitness is the scorecard in the inevitable competition that ensues; it is measured as a genotype's rate of increase relative to others in the population. For a lion, proficiency at smelling, running speed, and heat tolerance might all qualify as traits subject to natural selection. The first two might reasonably contribute to a lion's locating and capturing prey, while the third would reduce its dependence on water, all advantages for lions in the struggle to survive and reproduce.

Darwin was also among the first to discuss the importance of *sexual selection*, and he devoted an entire monograph to this subject, *The Descent of Man, and Selection in Relation to Sex*. Sexual selection refers to the "advantage which certain individuals have over others of the same sex and species solely in respect of reproduction" (Darwin, 1871). This advantage derives from intrasexual competition (typically male-male contests) or through intersexual choice (typically female choice among possible male mates). The scramble for reproduction, and hence sexual selection, tends to be asymmetric, because males often contribute little more to their offspring than sperm, while females often furnish offspring with energy reserves (e.g., the energy-rich yolk), protection, and sometimes extended nutrition and care.

As a consequence, females tend to be far choosier than males concerning their mates. Females allocate a majority of their resources to their offspring, whereas sexual selection on males leads them to allocate resources to traits and activities that enhance their success in fertilizing females. Females compete for areas with resources sufficient for raising their young, while males compete for areas offering access to females.* And because sexual selection commonly has a stronger effect on males, males vary more in reproductive success than do females—some males sire the majority of offspring, whereas others sire no offspring at all. Sexual selection is especially powerful in polygynous species, in which a single male monopolizes reproductive access to several or many females. Traits that help secure such fecund positions become exaggerated by selection, producing heightened sexual dimorphism. Sexual selection and polygyny are responsible for both elk antlers (via male-male combat) and peacock tails (via female choice).

Lion manes are obviously a sexually selected trait. Manes are restricted to males, contribute little to the lion's general economy of life, and apparently present a number of real costs that would otherwise make it a candidate for elimination by natural selection. Furthermore, manes are accompanied by sex-specific behaviors and pronounced sexual dimorphism—specifically the much greater size of male lions—and are often associated with a polygynous breeding system. Still, given the interest that people have focused on lions over the last thirty thousand years, it is remarkable that we do not know whether manes are the product of male-male contest or female choice, or possibly of both. Although typically treated as alternatives, these mechanisms are not mutually exclusive and may operate in concert. In the tropical American weevil *Brentus anchorago,* males have an elongated head that is subject to sexual selection: Males joust with these elongated snouts, and females mate preferentially with long-snouted males (Johnson 1982).

Competition among individuals of a given sex for mating opportunities is easily understood—any character promoting mating success is favored. However, sexual selection by intersexual selection—female choice—is far more complicated, because it involves characteristics borne by one sex and preferences for them carried by the other. Females initiate the process by selecting any trait that confers an initial survival or reproductive benefit.

*In lions, the territory of females represents a hunting ground, whereas for males, it is an area containing a number of females (Guggisberg 1975).

Even though traits and preferences may begin their evolutionary odyssey unlinked, they become genetically correlated over time as the offspring of such matings inherit both traits and preferences. So powerful is this genetic linkage between mating preference and sexual trait—whatever it may be—that it can even overcome substantial natural selection against the trait. This process has been called "runaway selection," because it is reined in only when reduced survival of the males outweighs the enormous reproductive advantage conferred by the trait. (For example, in the familiar cases of peacock and fancy guppy tails, sexual selection via mate choice actually opposes natural selection, which might otherwise favor more maneuverable and cryptic alternatives. Only the males inherit these sexually selected traits and their advantages or disadvantages.)

But mate choice can operate in several ways. An Israeli biologist, Zahavi, proposed the "handicap principle," which assays a male's overall background fitness by imposing an obvious handicap upon it (1975, 1977). Females should choose as mates those males able to survive and prosper despite this obvious burden. By this hypothesis, both sexes inherit the high overall fitness, but only the males express the costly handicap. American ecologists Astrid Kodric-Brown and her husband, Jim Brown, argued instead that sexual selection should favor traits that vary in parallel with overall genetic fitness. They developed the idea that selection favors the evolution of costly, phenotypically variable traits whose expression accurately portrays the survivorship and vigor of males. This "truth in advertising" model would permit females to assess accurately and directly the overall genetic quality of their potential mates. Traits fitting their description should exhibit little heritable variation, so that most of the variation among individuals reflects their age, conditioning, experience, parasite loads, and other proximate indicators of fitness. As in the handicap principle, traits functioning as signals of genetic constitution should also be costly to produce and maintain. This is essential to maintaining the honesty of the advertisement—it excludes would-be cheaters from hawking their inferior genetic constitutions with counterfeit labels of authenticity (Kodric-Brown & Brown 1984).

The sexual selection model proposed by evolutionary theorist Wirt Atmar takes the "truth in advertising" model one step further. Atmar sought to resolve that eternal conundrum—why should the females that provide most of an offspring's resources share parentage with free-loading males? Atmar used an engineer's viewpoint to argue that the overriding purpose of displays, courtship, and other interactions involved in mate choice is to expose underlying genetic errors and to purge them from future generations. Males are regarded in this model as the sacrificial caste that exposes genetic error to natural and sexual selection.* This is well known in many insects, in which the condition of haplodiploidy prevails. In this sex-determination mechanism, males have only a single set of each gene (making them haploid) rather than pairs of them (diploid), as females have. Consequently, any point mutation is fully expressed in males but buffered in females by a normal, unmodified copy. But even in exclusively diploid species (and these include most vertebrates), males often display a physiological fragility that is remarkable in view of their commonly greater physical size and strength. University of Washington anthropologist Don Grayson (1990) showed that it applies even to humans, determining that women fared better than men in surviving the extreme physiological ordeals of the Donner Party crossing. Metabolic exhaustion is a common theme in male-male contests among vertebrates, and the cessation of feeding and other maintenance behaviors during mating season only heightens the metabolic trials of male reproduction. A female can usually be assured that the last male standing is of a superior genetic constitution.

The manes of lions might fit any of these models of trait evolution. Manes result from the coordinated growth of guard hairs,† some up to 3.5 inches in length, on the face and atop the head, chest, and shoulders. Virtually all cats possess facial ruffs, and the mane appears to be an elaboration of this ruff onto the neck and shoulders. Manes extend for a variable distance down the nape, onto the shoulders, and onto the chest. They vary in color both individually and with age, from sandy or yellowish to tawny, chocolate, or black. Although the mane may afford protection in fights and is variable enough to permit individual recognition, its primary value is probably as a signal and general indicator of health and social status. The "lion strut," a behavioral display males make to subordinates, with legs stiffly stretched, head held back over the shoulders, chin down, and body presented in profile, likely evolved in concert with this ornament (Kingdon 1977).

Female lions might prefer males with bigger, flashier manes, as predicated by the runaway selection model. Manes may represent a fitness handicap, impairing hunting or maintenance activities in such a way that only high-quality males could afford to exhibit the appurtenance. Finally, manes

*Sexual selection is in fact one of a battery of such exposition mechanisms, an organism-level analogue of the DNA-repair mechanisms that operate at the molecular level.

†Two classes of hair cover most mammals: long, coarse "guard" hairs that shed water and often match substrates, and short, dense "underfur" whose principal function is to insulate.

may be a trait that embodies the underlying quality of a male's genetic constitution, signaling his fitness and underlying genetic quality to males and females alike. A brief review of patterns of mane variation serves to introduce some specific hypotheses that purport to explain why Tsavo's lions are maneless. These are taken up in Chapter 8.

Geographic Variation of Manes

Lions in Africa

Before humans precipitated its massive range reduction, lions occurred over vast areas of the Old World. In Africa, lions occurred from the Cape region to the Mediterranean coast and from the Atlantic to the Indian Ocean. They were absent only in the far reaches of the Sahara and the wettest areas of equatorial rainforest. Different living conditions throughout the range of *Panthera leo*—related to geography, climate, vegetation, and their variation—have led to differences among lions at the levels of individuals, habitats, and regions. The significance of this variation, which is strongly reflected in the size, coloration, spotting, and scantiness or profusion of the mane, is debatable.

The regional variation is striking. Dutch settlers extirpated the so-called Cape lions (*P. leo melanochaitus*) of South Africa during the middle of the last century. These lions were among the most distinctive of modern forms. Cape lions possessed a huge profusion of hair covering the crown, nape, neck, shoulders, and even some of the back. Often the mane was a deep black or blackish brown, sharply contrasting with the tawny flanks and accented only by a small yellowish-ochre fringe on the cheeks and forehead. The mane as a rule stopped short on the chest, so that the chest was devoid of mane hairs in the area of the sternum. However, most adult males developed a thick growth of long hairs running down the ventral midline to between the hind legs that could be called a "belly mane." Such appurtenances are very rarely seen in other African lions today, although Barbary lions also possessed them. Cape lions were distinguished from other African and captive lions by a tuft of hairs on each flank of the belly region (Mazák & Husson 1960). Cape lions also had broader and shorter, more bulldog-shaped skulls than lions to the north (Astley Maberly 1963).

The Barbary Coast of northwest Africa also supported profusely maned lions; like the Cape lion, these have also become extinct in modern times. Barbary lions had long, very thick manes extending to the middle of the back, and the manes were also thick and heavy on their underparts (Harper

1945). After studying all available material in European museums, Vratislav Mazák concluded that this was a highly differentiated race of lions that inhabited the entire Great and Little Atlas mountain system. Mazák claimed its range was absolutely isolated from those of other lion populations by surrounding deserts. Although sometimes regarded as huge, Barbary lions were about the same size as East African lions. However, they had darker and more grayish background coloration (resembling Cape lions in this regard), with long body hairs (1.2–1.4 inches) in both sexes. The large and very long manes reached behind the shoulders on the back; on the abdomen, two parallel strips of long hair formed the belly mane that thinned forward onto the chest. The same dark blackish color occurs in the mane, on the large elbow tufts, and in the belly mane. Although yellowish hairs surrounded the face, these graded imperceptibly into the dark mane covering the neck, shoulders, throat, and chest, so that no individuals exhibited the yellowish collar seen in Cape lions.

Late-nineteenth-century reports noted that "the Abessinian [*sic*] lion has only a short mane like that of Senegal" (Blanford 1870). But substantial geographic variation within Ethiopia exists: "The Ethiopian lion has a dark profuse mane which extends to the middle of the back and also underneath the belly" (von Wolffe 1955).* In color, extent, and particularly the belly mane, some Ethiopian lions rival those from the Cape and Barbary regions.

Frederick Selous wrote at length on the lions from Zambia, Zimbabwe, and Mozambique. In this region of Africa, manes are not huge, usually growing only around the neck and onto the chest and with a midline prolongation only from the nape to behind the shoulder blades. On occasion, however, Selous encountered large full-grown male lions that were practically maneless, as well as individuals whose entire shoulders as well as the neck were covered with mane (1908). It is noteworthy that Selous also noted parallel variation in their background coloration. On the Mababe Flats in Botswana, Selous shot two old male lions in excellent condition, lying together under the same bush. One was a full-maned lion with very dark-colored coat, whereas the other was a very light-colored animal with little or no mane (Selous 1881).

Ignoring such dramatic cases of individual variation, both mane and body color of African lions generally vary coincidently along a continuum. Desert-dwelling forms have lighter coats and yellower manes, while forest-dwelling lions are darker in both mane and pelage color, which can reach a dark tawny-yellow brown (Denis 1964). The characteristics of the lion of

*Both sources are quoted in (Hemmer 1974, 180).

Somalia, named by Lönnberg as *P. leo somaliensis*, offer a case in point. These lions' background coloration is very light. As in other East African lions, males are absolutely unspotted but on females very distinct dark yellowish spots on the belly and legs persist into adulthood. The mane is characteristically very poor, about 4 inches in length and confined to the neck. Its color is pale buff on the sides of the neck, grading to a blackish brown along a midline crest running down the nape. However, exceptions are commonplace: Former director of the Transvaal Zoological Gardens, Alwin Haagner referred to a "fine Black-maned Somali lion" siring two litters of cubs by a female lion originating in Congo (1920). Although Somali lions are on average smaller than those in other parts of Africa, Peel shot a male at Habr Heshi that measured nine feet three inches in length (Lönnberg 1912; Peel 1900). Citing historical sources, Charles Guggisberg documented parallel variation among the lions of Aïr, Niger, which also included both well-maned and practically maneless males (1961). Either two or more species of lions live side by side in many parts of Africa, or this species is hugely variable.

Some of the variation noted in lion manes rests in coloration, and cited references establish that mane color varies with geography and with age. Manes are also individually variable. As London zoologist R. I. Pocock qualified, "In my experience there is no such thing as a wholly 'black-maned' or a wholly 'tawny-maned' lion. Lions with the blackest mane always have the face surrounded by a tawny fringe.... On the other hand lions with the tawniest manes always show a certain amount of dark pigmentation along the median crest and low down in front of the base of the foreleg; there is very gradation between these types, and 'black' and 'tawny' appear to imply merely a preponderance of one colour or the other in the mane" (1931, 640). This color variation extends to individual hairs. In blackish manes, for example, most of the hairs are actually very dark brown, profusely mixed with black hairs, giving the mane a black appearance (Smithers 1983).

Elevational Variation

In Africa, as elsewhere, there is remarkable variation in many species with elevation. Air temperature invariably drops at higher elevations, as a consequence of the *adiabatic lapse rate*. The rate of temperature change with elevation depends on the water content of the air—its humidity—and varies from 3.3° F to 5.5° F per thousand feet. Common responses of warm-blooded animals to this physical variation are to minimize exposure to freezing temperatures at night and to grow longer, thicker coats with denser underfur at higher elevations. In lions, we see parallel variation in the length and thickness of the mane.

Harold Swayne was among the first to note this relationship for lions. He said, "Lions living in the Haud [on the Somali-Ethiopian frontier], where it is elevated five thousand to six thousand feet, have better coats and manes than those found in Guban or in Ogáden [below, in eastern Ethiopia], and the best skins I have seen have come from the elevated *ban* or open prairie. All the animals of the elevated country have thicker coats than those of the corresponding varieties found in the low country, this being necessary, no doubt, as a protection against the cold" (1895, 294). In Kenya, some of the finest lions on record are those from the Uasin Gishu, now the breadbasket of Kenya, an elevated area northwest of Nairobi. Among the many lions killed by Theodore Roosevelt and company on his famous Kenyan safari in 1909, one killed by Carl Akeley near Molo, Kenya (at nine thousand feet on the Mau summit), had the most black in the mane (Roosevelt & Heller 1914). Today, on the moorlands atop Kenya's Aberdares Mountains, older male lions have long, dark, dense manes that cover their heads, necks, and shoulders. Aberdares lions even sport tufts of long mane hairs on their flanks, a characteristic previously recorded only from Cape lions (Mazák & Husson 1960) but likely to characterize other high-elevation populations.

In *Man and Beast in Africa*, French naturalist François Sommer observed that lions at higher elevations seem to grow the finest and darkest manes (1954). Lions also grow large manes in the vicinity of large rivers, where continual evaporation causes cold nighttime temperatures (Astley Maberly 1963). In captivity, lions in European and North American zoos sport much larger manes than is typical for wild lions (Hollister 1917). Here, nutrition, condition and climate may all be involved. Although some have attributed this greater luxuriance in temperate zones as an indication that the mane evolved in Ice Age Europe, this seems unlikely. Heat loss in mammals varies as a function of surface area (the square of standard length*), while heat production varies as a function of volume (a cubic function of standard length). Given that males are already much larger than females and thus have a greater heat production-to-loss ratio, it is females, not males, that would be in greatest need of a special thermoregulatory coat.

Selous also contrasted the manes of wild and captive lions, stating in *A Hunter's Wanderings in Africa*, "I have never seen the skin of a wild lion with the mane equal in length to that attained by the greater part of the lions one sees in menageries." Fully maned lions in the wild have small tufts of

*An abstract combination of length, width and height

hair on the elbow and in the armpit but rarely or never have long hair along the belly and the forearm, which is common in European and American zoo animals. Selous characterized the coat of the wild lion as very short and pale yellow or silvery gray, unlike the thick reddish coats of captives, asserting that he "could pick out the skin of a menagerie lion from amongst a hundred wild ones." The differences were attributed to climate and nutrition. Given the prevalence of extensively maned lions in zoos and natural history books, many a sportsman was disappointed on seeing lions for the first time in Africa (Selous 1881, 259).

Recently, Tom Gnoske was quoted as having documented progressive mane length with elevation among lions in Tsavo West, from hot coastal areas to the cooler high plains (Anonymous 2001). Studies by Peyton West and Craig Packer also reinforce this coincident variation of lion mane length and elevation, attributing this pattern to the lions' tolerance of heat loads (2002). Lions in low, hot areas are near their limits of tolerance and sport sparse manes, while those in cool highlands feature longer, more lavish ones.

Lions in Asia and the Middle East

Lions are now restricted to a tiny fraction of their former range in Eurasia. Lions living in the Gir Forest reserve in northwest India possess a coat thicker than that of most African lions. They tend to have a longer tail tassel, larger elbow tufts, a more pronounced belly fringe, and a smaller mane than is typical among African lions (MacDonald 1984). Generally, the size of the elbow tufts correlates with the size and profusion of the mane, but in Indian lions, the tuft is large even though the mane is small (according to zoologist Nancy Neff in artist Guy Coheleach's *The Big Cats* [1982]). We do not know how many of these characteristics typified Eurasian lions generally and how many are idiosyncratic to the surviving Gir Forest population. That population has suffered founder effect, genetic drift, and allelic impoverishment since its turn-of-the-century bottleneck, when only a handful of Asiatic lions remained.

German evolutionist Helmut Hemmer argued that the Indian lion was an early derivative and current relict of a maneless lion stock that dated from the Pleistocene (1962). However, neither modern nor historic Indian lions are maneless. Adventurer and hunter Arthur Vernay stated "The black-maned lion has never been seen in India, the mane there being tawny, running to a very light yellow" (1930). In a comprehensive and definitive review of museum skins of Asian lions, the London Zoo's R. I. Pocock did not find a single maneless male among them (1931). Three of the four male Indian lions now preserved as tanned skins in the Natural History Museum, London, have more extensive manes than any of 15 male lions that Roland Kays and I documented in Tsavo East during our 1999 surveys (see Chapter 8).

Asian lions typically support a well-developed ruff around the neck, although on average this may differ in size and distribution from those seen in African forms. Friezes and sculptures throughout Asia Minor and the Middle East show that Persian and Mesopotamian lions had well-developed manes, even if few natural history collections from these regions document this morphology. One specimen from "Persia" in the Natural History Museum, London, had a far more extensive mane than "Romeo's" (shown being collared in Plates 10 and 11), extending on the midline to the rear of the ribcage and covering his entire neck and shoulders.

Another indication that lions from the Middle East were maned comes from captive rearing experiments. Lions from various zoos, including those of Jerusalem and Barcelona, were assembled on Longleat, an English manor, and reared together under uniform conditions. Lions of unknown parentage obtained from Jerusalem developed into well-maned forms that ranked among the park's most magnificent animals (Bath & Chipperfield 1969). *Contra* Hemmer, most authorities believe that the small, restricted manes of Gir Forest lions represent a secondary loss from a well-maned condition (Mazák 1964).

The Gir Forest is a dry deciduous forest growing at low elevations near the Indian Ocean coast. It is dominated by teak, whose importance in the forest community is maintained by forestry: logged areas are replanted with teak. The eastern part of the Gir receives only about twenty-six inches of rainfall, supporting an acacia thorn savanna, whereas rainfall increases in the west to about forty inches per year.* African lions living under these climatic conditions in Somalia, the Sahel, and the coastal scrub of East Africa also exhibit scanty manes. Thus, typically small mane size in Indian lions today is likely a dynamic response to modern-day environments there, not a remnant of their evolutionary past.

The Significance of Individual Variation of Manes

He was a magnificent beast, his mane covering the whole forepart of his body and even hiding his ears. Such fine manes are rare on wild lions, as most of them tear the hair out going through brush. (J. A. Hunter 1952, *Hunter,* 88)

*http://lynx.uio.no/catfolk/sp-accts.htm

While mane variation can be traced to various factors, anyone trying to summarize these patterns is obliged to issue disclaimers because of individual variability. True, lions at high latitudes and high elevations possess long, extensive, and richly colored manes, while those in equatorial climes and in hot dry regions possess smaller, yellower ones. However, the range of individual variation is so great that it is possible to find large, black-maned lions in virtually any part of Africa. Black-maned lions are not uncommon in the Kalahari, where they occur alongside those of the more normal tawny color (Smithers 1971). Similarly, Peel noted that the Somali lion typically has a very poor mane but that very rarely one finds a male with a large dark mane (1900). Such observations serve as reminders that any attempt to characterize geographic variation in a trait that is as individually variable as a lion's mane can hope only to present a statistical norm. To regard these differences as qualitative or categorical ones is almost certainly wrong.

Early South African settlers distinguished three, four, and even five species of lion, based upon the length and color of the coat and mane, spots on the underparts, and size. Austin Roberts distinguished three types of lions in southern Africa and allocated them to three different subspecies. The first, which he identified as *P. leo krugeri*, had a full dark or black mane and was found south of the Olifants River. The second, *P. leo vernayi*, had a yellow mane and was common in the Kalahari Desert, reaching into northern Kruger. The third, called *P. leo bleyenberghi*, the Katanga lion which ranged north into Congo, was pale yellowish gray in color and had a poorly developed mane, which extended from the ears only over the neck and with a short crest over the withers. But after spending time in Kruger, Charles Astley Maberly confessed that variation in color, body size, and development of the mane made it very difficult to assign any population to one of the three subspecies (1963).

Frederick Selous developed this perspective much more fully: "For my part, and judging from my own very limited experience of lions, I cannot see that there is any reason for supposing that more than one species exists, and as out of fifty male lion skins scarcely two will be found exactly alike in the colour and length of the mane, I think it would be as reasonable to suppose that there are twenty species as three. The fact is, that between the animal with hardly a vestige of a mane, and the far handsomer but much less common beast with a long flowing black mane, every possible intermediate variety may be found. This I say emphatically, after having seen a great many skins, and I entirely deny that three well-marked and consistent varieties exist" (1881, 257–258).

James Steven-Hamilton came to the same conclusion: "Although much has been written descriptive of different local races of lions, considerable observation of them in one area has led the writer, at least, to the belief that the seeming divergences of type exhibited in the presence or absence of mane, varying distinctness of spots in some adults, relative lengths of tail and so on, are in fact merely so many individual or family characteristics, dependent perhaps upon slight dissimilarities of environment" (1912, 165).

These authors recognized all extant African populations as *P. leo leo*, while others advocated a more complicated, heterogeneous picture. Roosevelt and Heller developed the case for regional subdivision. "Owing to the occurrence of black-maned lions occasionally in all the districts where lions occur, it is assumed that the color of the mane is of no racial value. We must not lose sight of the fact that in geographical races we are dealing with average characters and not absolutely distinctive ones, such as are possessed by species. The Cape and North African lions are usually black-maned, while the East African is decidedly a tawny or yellow-maned race. Black-maned lions are occasionally seen in East Africa, but they occur in the proportion of about one to fifty, and are of such rare occurrence that we are quite justified in calling the East African a yellow-maned race" (1914, 162).

Guggisberg dissented strongly: "Within the same geographical region there is so much individual variation that one cannot escape a feeling of considerable skepticism with regard to the ten to thirteen subspecies named by taxonomists, mainly on the basis of size, colouration of the coat, and colouration and development of the mane" (1975, 141). To serve usefully in diagnosing the different subspecies (or "distinct evolutionary units," as is more commonplace today), mane variation would need to correlate with other features of appearance or their underlying genetic constitutions.

Mane size and condition are not strongly correlated with body size. Stevenson-Hamilton noted that one of the best-maned lions he shot on South Africa's Sabi River was a comparatively small animal, while a far larger individual he took further north was practically maneless (1912). This independence of otherwise masculine traits also holds in the horn of Africa. Lord Wolverton reported two large male lions, measuring 10.8 and 10.6 feet in length, which were small-maned and black-maned, respectively (Guggisberg 1975). Both Selous and Astley Maberly noted evidence that both dark- and fair-maned lions may be born in the same litter and frequently consort with one another in the same environments (Selous 1881, Astley Maberly 1963). Manes fail to show the geographic consistency in development that we expect to characterize the distinct population clusters we recognize as subspecies.

When and How a Lion Develops Its Mane

The first signs of a developing mane appear at about a year of age, and shortly afterward male lions develop "sideburns" along the side of the neck and an erect crest of hairs that begins on the crown of the head and continues down the nape of the neck. Occasionally, elderly females develop shaggy sideburns that are faintly reminiscent of rudimentary manes (Guggisberg 1975). By two years of age, manes begin to grow conspicuous, extending around the front of the neck and onto the chest. Growth continues through the third year, and by the fourth, the mane forms a complete ring around the face. An isolated top-lock and bare patches behind the ears characterize this stage among Kruger lions. Finally, the mane reaches adult proportions after about five years, with the ears becoming inconspicuous and very long hairs growing on the throat and chest (Smuts, Anderson & Austin 1978).

Ned Hollister compared young lions nineteen to twenty-seven months in age that were captive in an area around Nairobi with wild adults lately killed in that region. Captivity produces a progressive darkening of the pelage, an increase in the length and thickness of mane and elbow tufts, and an extension of the mane over the base of the neck and top of the shoulders and along the belly, all independent of heredity. Consequently, even young adult captives rivaled the most impressive prime lions from the wild in terms of mane size and quality. Hollister also described the pervasive differences that captivity creates in skull morphology, mainly related to the greater use of jaw and neck muscles in wild animals (1917)

Writing of lions in Southern Africa, R. H. N. Smithers observed that the manes of subadult lions are always sandy, yellowish, or tawny. Only with advancing age do they sometimes become black (1983). Stevenson-Hamilton described the fifty largest male lions that he himself shot in what is now Kruger National Park, South Africa. Ten of twelve lions judged to be in the "early prime" of life had a "yellow mane," whereas two had "black and yellow manes." Besides these, he also examined twenty young male lions between eighteen months and 2.5 years of age—all had "incipient yellow manes" (1947, 343–345). Not all older males had black manes, but no younger males had them.

Mane development invariably begins with yellowish hairs. Guggisberg chronicled the subsequent development of a 4-year-old male lion in Nairobi National Park, Kenya. In July 1962, the lion had a scraggly yellow mane, but the mane had become fuller by October of that year and subsequently began to darken. By 1964 the lion had a very big yellow and brown mane, and by 1965, the mane was blackish brown with a broad yellow ruff (1975). Mazák confirmed that manes become darker with age, using case histories of captive animals at the Prague Zoological Gardens. Developing manes were always yellow or tawny, and only much later deepened to a dark color (1964).

Castration is known to inhibit mane development, as first reported by Pocock (1931). Guggisberg described another instance, involving a male lion in Nairobi National Park that had killed several females in inept attempts to mate with them. Because the park supported only a small population of lions, and this lion was then the most impressive cat in the park, authorities chose not to kill him. Instead, they castrated him to curb his amorous intent. Within three months, the lion had completely lost his impressive mane but resumed his attacks on lionesses (1975). This example highlights the dynamic role played by testosterone in mane growth and maintenance, a subject discussed further in Chapter 8.

Pleistocene Lions Were Maneless

Although first-hand reports and incontrovertible evidence are lacking, Pleistocene lions were probably maneless or practically so. In his book *Frozen Fauna of the Mammoth Steppe: The Story of Blue Babe,* Alaskan paleontologist Dale Guthrie reviewed various Paleolithic artworks and inferred that the Pleistocene lions of Europe lacked large contrastingly colored manes (1990). Instead, he argued, they wore discrete dorsal and ventral manes. Inconspicuous, minor manes like this matched Guthrie's theoretical considerations (reviewed in Chapter 8) and agreed with the scanty evidence on Pleistocene lions from the generally crude outlines and carvings of Ice Age artists. Recently, however, a treasure trove of Pleistocene paintings perhaps thirty thousand years old was found in a newly discovered cave in southern Europe, and these further clarify the matter.

The Chauvet Cave in the Ardèche River Valley of France contains superbly detailed paintings of various Pleistocene fauna. Lion depictions are everywhere; Craig Packer and Jean Clottes claim that "The seventy three representations of this animal here in Chauvet Cave exceed the total from all the other caves in Europe, and compared with these, all the previously discovered lion drawings are crude sketches" (2000, 57). The Chauvet paintings also appear far older, estimated at thirty-two thousand years old versus seventeen thousand for Lascaux. In them, large, more massive males and small, more gracile females interact in strikingly realistic poses. The paintings even depict what appears to be "hunkering," a component of courtship behavior (Packer & Cliottes 2000).

Many of the Chauvet lions are depicted in groups, as is typical of modern-day lions. These paintings imply that Pleistocene lions exhibited at least some degree of sociality. One important function of group living for lions lies in resource defense. Modern lions quickly consume prey, sometimes actively defending carcasses from the intense competitive pressures of scavengers. On the Mammoth Steppe of Pleistocene Eurasia, hyenas, wolverines, wolves, and polar foxes all scavenged from lion kills. By living in groups, cave lions could have reduced losses due to carcass theft (Vereshchagin & Baryshnikov 1992). However, primary productivity of Ice Age steppes in Eurasia and North America was far lower than in modern African savannas. And Pleistocene lions were even larger than their modern counterparts, increasing their per capita energy requirements. Lions range over significantly larger areas when prey densities decline during the dry season in Uganda (Van Orsdol, Hanby & Bygott 1982). These considerations imply that even small groups of Pleistocene lions would have needed vast territories to support them (Guthrie 1990).

George Jefferson studied Pleistocene American lions recovered from the Rancho La Brea "tar pit" deposits in southern California. He used the ratio of male to female size to infer their behavior and ecology. The sexual dimorphism of *P. leo atrox* (26 percent difference in femoral estimates and 31 percent in dental estimates) was less than that of tigers (>40 percent in *P. tigris*) and greater than that of modern lions (about 15 percent in *P. leo leo*). In addition, the La Brea sample of *P. leo atrox* contained roughly equal numbers of remains from adult males and females, whereas modern African populations are predominantly adult females. As a result, it is unlikely that Pleistocene North American lions lived in prides like modern African lions or hunted cooperatively in social groups. Jefferson's findings agreed with Guthrie's analysis of European cave lions—both appear to have hunted in pairs or alone (Jefferson 1992, 104–105). Perhaps the Chauvet Cave was in an area of particularly rich game concentrations or represented an unusually productive interval between glacial advances.

Remarkably, none of Chauvet's seventy-three depictions of lions shows any trace of a mane, leading Craig Packer and Jean Clottes to consider them "maneless pride dwelling lions." Did the Pleistocene artists of Chauvet choose to paint only lionesses or neglect to include manes (Sutcliffe 1985)? This seems far-fetched, given their otherwise eloquent depictions of even subtle features of morphology and behavior, such as tufted tails, scrotal sacs, and "hunkering." And modern artists almost always portray a heavily maned male lion. Pleistocene artists may have been less chauvinistic than their successors, but I wouldn't bet on it. All evidence points toward Ice Age lions' lacking manes, which makes the prevalence of this trait among modern lions only ten millennia after its close all the more remarkable.

Chapter 8
Why the Lions of Tsavo Are Maneless

... after all, what is a lion without a mane but the shadow of the noble beast one has mentally pictured to oneself? (F. C. Selous, *A Hunter's Wanderings in Africa*, 1881, 259)

The earliest known references to maneless lions are in Roman texts. The insatiable appetite of Romans for gladiatorial sport led them to scour countries near and far for suitable combatants. Initially, the chief supply of lions for these events would have been from Morocco and Algeria, where the Barbary lion lived. These, of course, bore very heavy manes, but many of the lions obtained by the Romans were maneless (see Plate 7). Various ancient authors carefully distinguished maned and maneless lions; for example, the Emperor Probus tallied one hundred Libyan and one hundred Syrian maneless lions and one hundred lionesses. Obviously, this author distinguished males and females.* Because maned and maneless lions were obtained from the same regions, they were unlikely to have been distinct geographical races. Because they were intended for blood sport at gladiatorial games, the maneless lions could hardly have been juveniles and still satisfied the audiences. Historian W. M. S. Russell even suggested that this must represent a "transient polymorphism" in lion coat variation (1994).

Noted cat authority Paul Leyhausen disputed the existence of distinct races of maneless lions, either in Roman times or today. He argued that the sustained intensity of the Roman harvest, which spanned decades and even centuries, probably resulted in the capture of large numbers of younger

*Pliny knew that lions grew manes only upon reaching sexual maturity, but so essential were manes to his conception of lions that he believed maneless lions had been sired by leopards (Russell 1994).

lions. Subadult males commonly reach adult body size before their manes are fully grown, and these might have been distinguished by Roman chroniclers. Leyhausen argued that the Romans used "maneless" as merely a descriptive rather than a classificatory term (1995). Like many other authors, Leyhausen discounted reports of adult maneless males.

Roosevelt and Heller were among the first to attribute manelessness categorically to age. They noted that it is impossible to accurately determine the age of a lion in the field without knowing his personal history. "Nothing, for instance, is more common in the literature than the statement of the occurrence of adult maneless lions. A careful examination of many museum specimens, however, has failed to find really old lions ... associated with maneless skins. The so-called maneless lions are really immature specimens of adult size which have every appearance of being fully adult and are on that account considered so by sportsmen.* Adult size of skull or body is not a reliable character of maturity; immature animals not only equal but occasionally exceed the mature ones" (1914, 162–163). Charles Guggisberg also traced reports of manelessness in Indian lions and in lions from various parts of Africa to instances in which subadult animals had been mistaken as adults (1975).

The hypothesis that maneless lions are invariably immature is relatively easy to disprove. In a technical presentation to the American Society of Mammalogists in 1999, my colleagues and I plotted mane size against age for a sample of African lions at the Field Museum represented by both tanned skins and cleaned skulls. Age was determined by tooth wear and the degree to which various cranial sutures had closed, a method worked out by Smuts and colleagues in a large sample of lions of known age (Smuts, Anderson & Austin 1978). Manes were scored subjectively from 1 to 10, where 1 represented no mane growth whatsoever and 10 represented the long, luxuriant growth typically found only in zoo animals. The results (Figure 11) showed several young animals that had not yet grown manes, as well as older lions with long manes. Together, these points define a normal "growth curve" for mane development over time. However, a large number of old—in some cases, very old—individuals had manes no larger than those borne by immature lions; these include

*Curiously, Roosevelt and Heller also committed these same errors! Elsewhere in the same volume, they described maneless lions without qualifying reference to age. Writing of two males shot on the Athi Plains in company with a lioness and cub: "They were two big fine fellows, but as customary with Athi lions, almost maneless" (1914, 208). Later, he wrote of one shot on the nearby Loita Plains, "He was a full grown male, but maneless, and in excellent condition, with a heavy layer of fat under his skin" (216–217).

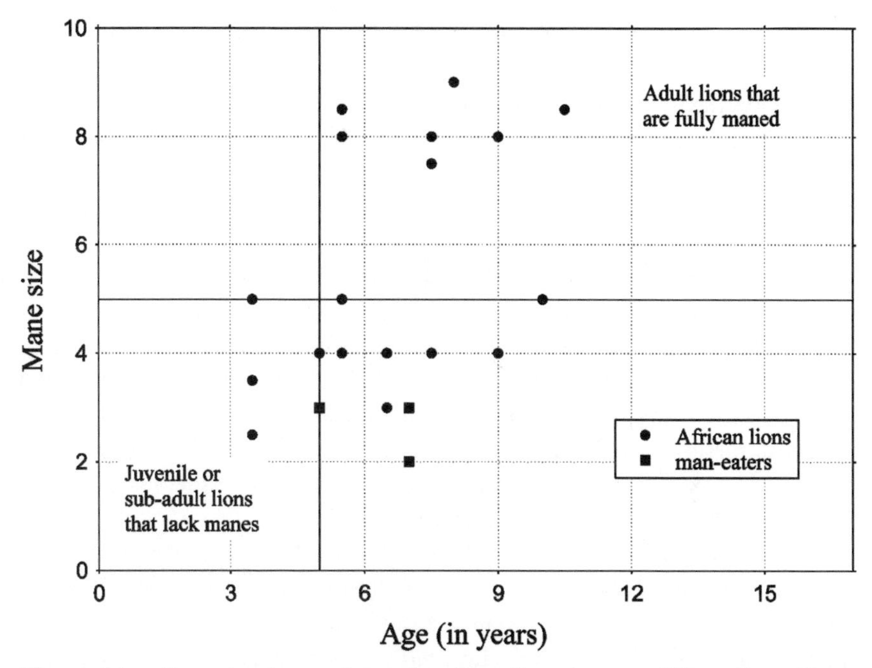

Figure 11. Mane development as a function of age in some African lions. Manes were scored on length, extent, and color from 1-10, while age determination employed tooth-wear and suture-closure criteria developed by Smuts et al. (1978). Lions normally have a well developed mane (score ≥5) by 4-5 years of age; the large number of older but scanty-maned lions in the lower right quadrant documents the existence of manelessness (graph from Patterson et al., 1999).

the historic Tsavo man-eaters. Obviously, sexual maturation was not accompanied by mane development in these lions (all considered by taxonomists to be *Panthera leo leo*).

Three of the individuals with the scantiest manes were the two Tsavo man-eaters and the man-eater of Mfuwe. Lions in the low-lying scrub along the Indian Ocean coast from Somalia to Mozambique have notoriously poor manes. Some writers have even described them as utterly maneless. Even longtime denizens of this arid scrubland differ in their opinions on the existence and prevalence of maneless lions. John Hunter, who spent most of his time in the bush in Kenya and Tanzania, stated, "Maneless lions have been recorded in a number of locations in Africa (including Hwange), but they are fairly unusual. In most cases, the animals probably once had a mane but then lost it, usually as a result of injury, poor condition, or old age" (1952, 26). On the other hand, John Taylor noted that, in the Macua stretch of northern Mozambique, lions inhabiting the dense thorn bush are utterly maneless. He noted that even if they had once had manes, the hairs would soon have been ripped to pieces in the bush (1959).

In 1998, I traveled to Tsavo in the company of Chap Kusimba, Tom Gnoske, and Julian Kerbis Peterhans to determine whether maneless lions still roamed Tsavo National Parks. We saw only three male lions on that trip, but two were certainly adult and all three were as maneless as the historic man-eaters. The following year, Roland Kays and I spent four months in Tsavo East National Park, surveying and identifying lions in the area between the Galana and Voi rivers. Our exhaustive inventory in this region eventually documented eighty-seven individual lions with repeated observations. None of the males we observed could be called well maned—even the most hirsute would be considered scantily maned in other parts of Africa, and some were virtually without manes. Manelessness is a fact in Tsavo lions, but how do we account for it?

In the remainder of this chapter, I draw extensively on my productive collaboration with Roland Kays, now the Curator of Mammals and Birds at the New York State Museum. Many of the ideas I discuss here were developed jointly with Roland while he held a postdoctoral fellowship at the Field Museum, supported by the Barbara E. Brown Fund for Mammal Research. Our discussions of these ideas are so numerous and extensive that it would be difficult for us to separate our progress or understanding of these issues. Since 2001, we have also been collaborating with Samuel Kasiki of the Kenya Wildlife Service, who likewise is a full partner in the ideas and analyses discussed below.

Nature vs. Nurture: Hypotheses to Explain Manelessness

Any complex biological trait may be influenced by heredity and environment. Often, detailed studies reveal that both effects are involved—that neither "nature" (genetics) nor "nurture" (environment) exerts definitive effects. Rather, heredity and environment typically interact to produce the observed distribution of the trait. This is a common conclusion in studies of human intelligence. Still, "nature" and "nurture" constitute a useful way to classify several hypotheses that have been proposed to explain manelessness. In this chapter, I distinguish *evolutionary hypotheses,* which propose underlying genetic changes affecting manelessness, from *developmental hypotheses,* in which manelessness develops (or doesn't) over the course of an individual's growth and maturation, in the absence of genetic change. Although my discussion treats evolutionary and developmental hypotheses as distinct and specifies causation in such a way as to render them discrete and nonoverlapping, the reader should be aware that they are nonexclusive and may operate in unison.

The ultimate test for distinguishing between genetic and environmental causes of manelessness would be a simple captive-rearing experiment: Raise Tsavo males in a controlled environment, such as a zoo, where lions ordinarily develop a mane. If Tsavo males failed to develop a mane under these circumstances, then we would have eliminated environment and nutrition as possible causes and could focus on genetic causes. If, on the other hand, Tsavo lions developed a mane in captivity, then nutrition and environment would gain credence as causal factors. If this experiment has been conducted during the course of animal management,* records of its outcomes have not been kept.

Evolutionary Hypotheses

The "Basal Lion" Hypothesis

The most exciting possibility raised by the existence of maneless lions was that they might represent a distinct species or subspecies of lion. It was this possibility that initially attracted me to work in Tsavo and led me to collect nearly a hundred tissue samples from Kenyan lions in 1998 for DNA sequence analyses (more on this shortly).

We know with certainty that within the Felidae, manes are restricted to lions—no other cat has them. This tells us that the lion ancestor lacked a mane. Moreover, as we have seen, Pleistocene lions were fundamentally maneless, so perhaps Tsavo's lions represent a descendant or close relative of this primitive lion! Thus, maneless lions might represent a basal lineage of lion that never had a mane, possibly one showing a special relationship to Pleistocene lions. This is the interpretation favored by Tom Gnoske and Julian Kerbis Peterhans (Gnoske & Kerbis Peterhans 2000; von Buol 2000). Their idea revives suspicions that there may be unrecognized variation in African lions, perhaps even including an unrecognized species. Another, earlier example comes from the Kalahari, where a supposedly different species of lion was thought to inhabit the Kaokoveld portion of the Kalahari desert—residents there described a smaller mountain race of maneless lions that were more ferocious and less social than lions in the Kalahari proper (Eloff 1973).

*The Kenya Wildlife Service often employs translocations of problem animals as an alternative to euthanasia and sometimes collects orphaned youngsters of various species to be reared at "The Orphanage," located at the gate of Nairobi National Park. Both courses of action—transplanting Tsavo males elsewhere and introducing maned lions to Tsavo—would generate observations of interest with respect to manelessness.

But the maneless condition might alternatively be derived—maneless lions might represent a distinct lineage of lions that has secondarily lost what is otherwise now a species-wide trait. Under both scenarios, maneless lions would be genetically distinct from other, maned lion populations. But in the first alternative, lions would lack the gene responsible for the production of manes, while in the second, that gene would be present but somehow lost, incapacitated, or switched off. Genetic divergence between maned and maneless groups would be quite strong under the "basal lineage" hypothesis but might be relatively modest under the "secondary loss" scenario.

To test these hypotheses, my colleagues and I evaluated the characteristics that mammalogists usually employ to gauge evolutionary distinctiveness. We gathered cranial, dental, and genetic data on both maned and maneless lions. If either hypothesis is true, then we should find significant differences between maned and maneless populations in one or more traits to document this evolutionary distinction. Lack of differentiation in both morphology and genetics would indicate that little or no appreciable evolutionary divergence separates maneless and maned populations of lions.

Figure 9 (page 114) showed a scatter plot of 176 male and female African lions, all currently considered *P. leo leo.* Data points in the figure represent lions from eastern and southern Africa, because maned and maneless populations occur in close geographic proximity throughout these regions, which eliminates regional variation as an uncontrolled variable. The two variables used are the greatest breadth of the skull (measured across the cheekbones) plotted against its greatest length. Jointly, these variables depict overall skull shape. Samples from Tsavo East National Park (closed diamonds) do not fall outside the range of variation shown by other eastern and southern African lions (open circles). They are no larger, smaller, broader, or slenderer than other lions. In fact, Tsavo samples encompass much of the variation seen in lions elsewhere in Africa. The historic Tsavo man-eaters and the man-eater of Mfuwe fall smack in the middle of the modern sample from Tsavo.

The plot in Figure 9 shows relationships among lions in only two variables, skull length and cheekbone breadth, but various other pairs of variables produce highly similar plots. The fundamental similarity of Tsavo lions and other populations from East Africa is also shown with a multivariate tool called *principal components analysis,* which simultaneously considers similarities and differences across dozens of variables. Here again, Tsavo lions cluster in the midst of a large cloud of variation among East African lions, and these broadly overlap similar clouds seen in other parts of Africa (Patterson et al. 1999).

An even stronger refutation of the "basal lion" hypothesis is provided by DNA sequence analyses (Dubach et al., in press). Jean Dubach, a conservation geneticist at Brookfield Zoo's Rice Conservation Biology and Research Center outside Chicago, led a team analyzing genetic samples from lions throughout eastern and southern Africa. Included in the analysis were Tsavo samples from three supposedly different "types" of lions recognized by rural people in East Africa: big-maned "pride lions" typical of the Serengeti or Masai Mara, big maneless lions denoted as "buffalo lions" by Gnoske and Kerbis Peterhans, and small maneless lions called *simba marara,* much feared in Kenya and northern Tanzania as stock raiders and potential man-eaters. DNA of Tsavo lions was extracted from dried skin samples and compared with DNA samples taken from lions elsewhere. Our initial presentation (Patterson et al. 1999) analyzed variation between samples over twenty-three hundred base pairs of DNA, encoding two mitochondrial genes (cytochrome b and nitrogen dehydrogenase subunits 5 and 6). Dubach's preliminary analyses showed that lions differ from tigers in hundreds of these base-pairs. Although these genes probably have no role in the production of manes, they are useful as genetic markers. Given the immense size of animal genomes (literally billions of nucleotides make up the human genome), scientists commonly select a small genetic sequence thought to be representative of the genome as a whole, document its variation, and then extrapolate sample similarities and differences to the entire genome.

The results were striking: All five Tsavo samples representing the three "morphs," or body types, had identical base-pair sequences for these genes! Not a single point mutation separated samples of maneless scrub lions living near Voi from that of the well-maned lion we included from the high plains of Tsavo West. Equally striking was the fact that this *identical* gene sequence was also detected in lions from as far away as the Transvaal of South Africa, yet different South African populations exhibited dozens of differences. Not only do maneless lions lack a gene sequence distinguishing them from other East African lions, but East African lions as a group are poorly distinguished from South African lions. Finally, the Tsavo-Transvaal group was not basal (diverging prior to the splits of other lions) but rather was a later, derived lineage. An inescapable conclusion of this analysis is that whatever evolution separates maneless lions in Tsavo from other African lion populations has been relatively recent.

Together, the morphological and genetic analyses constitute a convincing refutation of the "basal lion" hypothesis. The essential identity of Tsavo lions, both among themselves and with other African lions, suggests that only a modest amount of independent evolution can have taken place since they shared a common ancestor with other African lions. Because their relatives and other more ancestral lions are all maned, Tsavo lions must have secondarily *lost* their manes, rather than never having had them in the first place.

Dubach and colleagues are continuing these surveys of genetic variation in lions, exploring variation in faster-evolving regions of the genome. Dubach has determined that a highly variable region of the mitochondrial genome known as the "D-loop," or control region, varies among Tsavo lions and may offer a means of identifying maternal lineages. Hypervariable "microsatellite" regions in the nuclear genome offer additional means of resolving lineages. It is too early to say whether maned and maneless lions can be differentiated on the basis of D-loop sequences or nuclear microsatellites. If so, this would elevate the possibility that manelessness has a genetic cause. Alternatively, if manelessness were caused by environmental influences, no genetic differences of any sort would be expected.

The "Costly Mane" Hypothesis

Another hypothesis states that lions should lack manes when the advantages conferred by manes are offset by the disadvantages they pose. Manes are supposed to confer social advantages, making their possessors more attractive to females and more intimidating to males. Perhaps manes also make lions more intimidating to females, as females often ferociously reject strangers in their territories. On the other hand, manes might entail diminished hunting success via conspicuousness, meaning that prey might detect hunting lions with manes at greater distances and so effect their escape. Manes could also limit the maneuverability of lions hunting in dense brush, slowing their chase of prey fleeing through thickets or dense grass. The state of mane development should represent a dynamic compromise between social benefits and environmental costs.

Prides of females are the basic unit of lion social organization. Opportunities for breeding with females reside mainly within a pride structure, creating competition among males for access to them. When females are banded into large groups, competition among males is especially severe and the disparity between "haves" and "have-nots" especially great. Frequently, male lions gain access to females as part of a coalition, a partnership of males set up to acquire and maintain those reproductive privileges. Often, but not invariably, coalitions are composed of brothers and cousins. The scientific concepts of "inclusive fitness" and "kin selection" may help to explain the apparently selfless behavior of one lion defending another's access to a female in estrus—because he shares genes with the breeding male, the defender's genes are also being propagated.*

Lions are highly intelligent animals. Playback experiments using recordings of lion choruses have shown that lionesses can accurately assess the numbers of callers in broadcast recordings—they choose to press territorial claims or to retreat based on the perceived number of challengers (Heinsohn & Packer 1995). Other experiments have simulated the presence of intruder males and monitored the responses of residents. Resident males invariably approached the broadcast roars whenever two or more males were present and failed to approach the intruders only when a single male faced the roars of three intruders (Grinnell, Packer & Pusey 1995). Lions use this acoustic information to formulate clear-cut strategies of territorial and self-protection.

Experiments on the role of visual symbols in lion communication are only beginning, but there can be little doubt that manes serve to advertise the presence and condition of males. Overall pelage coloration of lions is remarkably effective as camouflage, being lighter on sandy-colored soils and more richly colored elsewhere. At dawn or dusk, a group of lions can virtually dissolve into rocks, soil, or bushes simply by becoming immobile. The coloration of the ears, the tail tuft, and (in males) the mane contrasts sharply with the background color of lions. In the former two, this serves to accent the position and motion of their body parts, possibly adding emphasis to visual signals, as does the contrasting color of wolf lips. Manes also serve as a whole-body accent, making the lion's head, neck, and chest appear larger than they otherwise would. The complexity of the mane's distribution over the body (cheek, nape, forelock, chest, neck, shoulders, back, elbow tuft, etc.), as well as variation and contrasts in the length and color of its hairs, create a complex signal that is potentially rich with information for other lions.

Manes contribute to impressions of size, and size is perhaps the most widely used arbiter of dominance in the animal kingdom. By enhancing a lion's outline as seen from the front and side, a mane enhances his apparent

*Altruism and apparently selfless acts present a challenge to evolutionary theory, in that selection acting on individuals should eliminate such incipient developments and replace them with purely selfish alternatives. However, a wide variety of studies have shown that by aiding the reproductive efforts of kin—whether parents, siblings, or children—organisms can increase their overall ("inclusive") fitness, even at the expense of their own individual fitness. The rationale behind this arithmetic was made famous by the British geneticist J. B. S. Haldane, who once was asked if he would risk drowning to save his own brother. "No," replied Haldane, "but I would to save two brothers or eight cousins." http://www.abc.net.au/science/descent/trans1c.htm

dominance, which may be useful in averting costly, potentially damaging conflicts. The mane also differentially exaggerates traits whose development differs between the sexes by festooning and augmenting the head, chest, and shoulders, features already enlarged by sexual dimorphism. The mane may serve to exaggerate the visual signal of a lion's masculinity, and do so in a manner that both females and males can read. Alternatively, manes may protect precisely those areas most vulnerable to injury in conflict with other lions: the nape, throat, and shoulders.

University of Alaska paleontologist Dale Guthrie proposed the most explicit hypothesis on the evolution of manes to explain manelessness in Pleistocene lions (1990). He argued that low prey densities on the Ice Age steppe would have limited lions to small groups, effectively two or three individuals and their cubs. Because many small groups of females would result, mating opportunities could not be monopolized by one or a few males. As more males contributed to the next generation, the strength of sexual selection among them was reduced.* Meanwhile, the number of lions cooperating in hunts fell to the point that male participation in hunts was crucial to the group's fortunes—the very large size of Pleistocene herbivores only underscored the value of male hunting abilities. And because manes are conspicuous and cumbersome, they could only hinder the hunting activities of males.† Thus, Guthrie hypothesized that there would be natural selection *against* manes, even as sexual selection *for* them virtually disappeared. If lions had already evolved manes by

*The strength of sexual selection is commonly measured as the variance in male reproductive success. The greater the disparity between the "haves" and the "have-nots," the stronger sexual selection will be, enhancing selection pressure for sexually selected traits. Among mammals of the African savanna, this reaches its zenith in the impala, *Aepyceros melampus;* males commonly tend harems of fifty to eighty females.

†Roosevelt and Heller doubted that manes hindered hunting activities in any way. They stated, "The male lion has some strongly revealing bodily attributes. His mane is conspicuous, and when it is black it has a highly revealing quality, yet the black-maned lions are generally beasts in high condition; apparently neither the presence of this highly revealing black mane in some males, nor the absence of all mane in the females, has any effect one way or the other in helping or hampering the animal against its prey.... The lion's coloration is really a wholly minor, and perhaps a wholly negligible, element enabling it to approach its prey unperceived—in other words, that the undoubtedly concealing quality of the lion's coloration is of interest chiefly as a coloristic fact, and plays little real part, and probably no part at all, in the animal's success as a hunter, and has not been developed by natural selection or otherwise for this particular utilitarian purpose" (1914, 172–173). This contrasts, however, with the impressions lions leave on other naturalists. J. A. Hunter said, "A lion's skill in concealment is extraordinary. I have seen a big lion crouch low and take cover behind some grass that I thought would scarcely hide a hare" (1952, 43).

the end of the Ice Age, nature's economy would quickly eliminate this needless and costly luxury (much as it has done with the eyes of blind cave fishes and salamanders).

Guthrie even cited Tsavo's lions as support for his hypothesis, saying, "In areas of Africa where prides are quite small, male lions today have small, noncontrasting manes. Both in the heavy bush of Tsavo Park (Schaller 1972) where there is low game density and in the Kalahari (pers. obs) where prides disperse during the dry season for a more solitary life, lions have small tan manes" (1990, 105–106). Guthrie acknowledged that African lions differ from Pleistocene lions in various ways that might disrupt the analogy, including higher density, reproduction throughout the year, and typically short tenure of pride males. Nevertheless, he succeeded in assembling the first coherent hypothesis for manelessness based on its selective cost.

In 1999, Roland Kays and I set out to test this hypothesis by acquiring new data from the field (Kays & Patterson 2002). We conducted surveys of lions in Tsavo East and Tsavo West from September (dry season) through December (which was quite wet). We used a vehicle to search Tsavo roadways for recent lion tracks, then broadcast the tape-recorded call of an injured buffalo calf over a public-address system to attract groups of lions into view, identifying the arriving individuals by age, sex, and condition. We had repeated encounters with most of the lions that reside in Tsavo East south of the Galana River, and we were able to verify their social groups on multiple occasions. Because this was the first systematic search for lions that explicitly considered the condition of manes, it was also the first definite proof that the lion population in the southern part of Tsavo East was predominantly maneless. As our report was published, *National Geographic* featured an article on Tsavo's maneless lions written by Phil Caputo, providing superb photos of maneless lions there (2002).

Although we expected to find that maneless lions were prevalent in Tsavo, the survey's other findings were surprising (Kays & Patterson 2002). Park surveys in the 1970s had shown that Tsavo supported low densities of lion prey, and (like Guthrie) we expected this would translate into small pride size. However, the five prides we studied (in the vicinities of Aruba, Kanderi, Sala, Satao, and Voi) averaged 7.4 females per pride. This is actually slightly *larger* than the average number (6.5) of females that George Schaller documented in fourteen Serengeti prides (1972).

With Tsavo females even more clumped than is typical for lions in the high plains, the effect of sexual selection on males should be stronger, not weaker, and result in more exaggerated manes rather than less developed ones. Even more striking was the fact that a single adult male attended each of our five prides, and that the male was maneless or practically so (Plate 8). The fourteen prides that Schaller documented in the Serengeti were each tended by a coalition of 2–4 males—in no case did a lone male defend a Serengeti pride. Death or injury to a coalition member frequently resulted in expulsion, injury, or death for his partners.

Thus, in Tsavo we found what appears to be a novel social system for lions: large groups of females tended by a single maneless or nearly maneless male. The absence of male coalitions in Tsavo apparently is not a consequence of a sex ratio more skewed toward females, because nomads in the park were predominantly males without prides. Curiously, we frequently documented nomadic groups containing 3–4 males within a day's travel from single-male prides (Kays & Patterson 2002). We do not know how resident males are able, by themselves, to defend their prides from takeovers by these larger groups.

Whatever the explanation for the social systems we documented in Tsavo, Guthrie's hypothesis for manelessness is refuted—Tsavo's lions are maneless, but the scenario for sexual selection is undiminished as females occur in large groups. It is noteworthy that the social behavior of extinct Barbary lions also refutes Guthrie's hypothesis. Those lions had perhaps the most extensive, strongly contrasting, and elaborate manes known. However, Barbary lions occupied mainly wooded areas and lacked a pride social structure. Except for females with young, they were strictly solitary. Perhaps sexual competition among males was intensified by highly seasonal mating habits of these lions: Rather than reproducing year round as do most other lions, Barbary lions bred in January (Mazák 1970). In any event, the existence of highly maned Barbary lions without prides and maneless Tsavo lions with large prides effectively refutes Guthrie's hypothesis of selection based on group size.

Why doesn't the group size of Tsavo's lions correlate with prey density as it does in Uganda (Van Orsdol 1984)? Despite low *average* prey densities in Tsavo (Leuthold & Leuthold 1976), the region's pronounced dry season may effectively concentrate prey at *higher* densities around permanent water sources; most of our social data were collected during this period. Perhaps the average prey of Tsavo's lions is larger, stronger or faster, requiring more cooperating hunters. Other explanations for large group size in lions (protection against infanticidal males and predatory or carcass-thieving hyenas (Packer, Scheel & Pusey 1990) seem less likely but cannot be discounted. Regardless, because 5 male lions effectively controlled reproduction with 37 pride females, and these represented a majority of breeding lions in southern Tsavo East, the strength of sexual selection on Tsavo males should be even higher than it is in the Serengeti or Masai Mara. However, this observation refutes only one version of the "costly mane" hypothesis.

Temperature and Water Costs of Manes

Alternatively, manes might heighten the energy and water costs of thermoregulation for male lions. This variant of the "costly mane" hypothesis holds that the costs imposed by manes are mediated by the lion's water economy. By interfering with heat loss, manes could increase a male's water requirements. To maintain its water balance, a male so encumbered would be limited to a smaller hunting radius around permanent water sources (at least during the dry season, when water is scarce). Alternatively, his hunting and traveling activities might be restricted to the cooler twilight and nighttime hours rather than extending into daylight. This hypothesis has not been fully tested. Roland Kays, Samuel Kasiki, and I are now collecting data on daily and seasonal variation in time budgets and ranging behavior, with special reference to temperature and water. Nevertheless, certain observations are possible.

Mammals and birds are both "warm-blooded" (homeothermic), with body temperatures generally higher than their surroundings. Both groups generate the heat necessary for elevated body temperatures by revving up their metabolisms. Both retain this heat by means of their epidermal derivatives, hair in the case of mammals and feathers in the case of birds. Prodigious consumption of energy to power their metabolic furnaces is an integral part of the physiology of homeotherms. In fact, it takes special adaptations, like those involved in hibernation or estivation, to suspend the continual production of heat. (Both of those involve extended periods of inactivity.)

It is well accepted that mammals more easily tolerate environmental temperatures *below* their zone of "thermoneutrality" (where constant body temperatures are possible without elevated metabolic rates) than when temperatures rise *above* it. A variety of energetically inexpensive behavioral and physiological responses to cold temperatures are possible: tolerating hypothermia, erection of the hair, huddling, and reducing blood flow to the body surface and extremities. Even shivering represents merely an extension of the basic mammalian strategy.

However, *hot* temperatures are another matter. Many species use behavior to reduce temperatures, shifting their times of activity, den sites, and the like to minimize heat loads. In addition, most use evaporative cooling of water to regulate their body temperature, either by panting or by sweating. Some highly adapted desert dwellers, such as camels and oryx, are famous for their tolerance of hyperthermia. In these animals, body temperature may climb to 113° F, which would be lethal in many other species. Such tolerances rely on a reorganized circulatory control system that shields the brain from overheated blood (Vaughan, Ryan & Czaplewski 2000).

Tsavo is famous for being both hot and dry, and Roland Kays and I proposed that these factors create powerful selection pressures against large manes in its lions (Kays & Patterson 2002). Every animal in Tsavo experiences heat and water stress, at least seasonally. Species may shift times of activity, habitat preferences, food habits, social groupings, or shelter requirements in response to the region's special demands. As a trait with largely social consequences, manes may represent a nonessential appurtenance that is too costly in metabolic terms to maintain. Hence, any genetic variation in the ancestral population would under these circumstances be subject to directional selection for the reduction and perhaps even the elimination of manes. By reducing the heat loads and associated water costs of lions possessing it, the maneless condition may permit longer foraging times (especially in daylight) and larger foraging radii and territorial patrols around dependable water sources.

If this hypothesis were true, then the distribution of maneless lions should coincide with hot, dry environments. Accordingly, I sought mapped contours for temperature (as a proxy of heat stress) and precipitation (as a proxy for potential water stress). Wijngaarden and van Engelen explored the climates of the greater Tsavo ecosystem in their book *Soils and Vegetation of the Tsavo Area* (1989). They were keenly aware of the complexity of biological responses to aridity and mapped various combinations of complex response variables. One was the probability that the rainfall during both the long rains (March to May) and the short rains (October to December) will be less than 90 percent of the potential evaporation. Meaningfully, about half of Tsavo East National Park stands more than a 90 percent chance of receiving half as much rainfall as evaporation, and at least three-quarters of it stands at least an 80 percent chance (van Wijngaarden & van Engelen 1989). This is a dry, parching environment. These authors' maps of average annual precipitation show that about 60 percent of Tsavo East National Park receives less than 300 mm (twelve inches) of rain, and no area south of the Galana River receives more than 500 mm (twenty inches). In contrast, except for an arid corridor running along the Tsavo River, most of Tsavo West receives at least 400 mm (sixteen inches) of rain annually. The 400-mm contour on the map (Figure 12) approximates the distributional limits of maneless lions we have documented in Tsavo.

Surrounding areas generally receive more precipitation, and these may be the source of the well-maned lions that are occasionally seen in the parks. For example, Galana Ranch (on the eastern border of Tsavo East) lies in a coastal climatic zone that receives sixteen to twenty-four inches of rain; lions from this zone may periodically turn up near Buchuma Gate or Lake Jipe. The decently maned lion we have radio-collared on Taita Ranch (Plate 4),

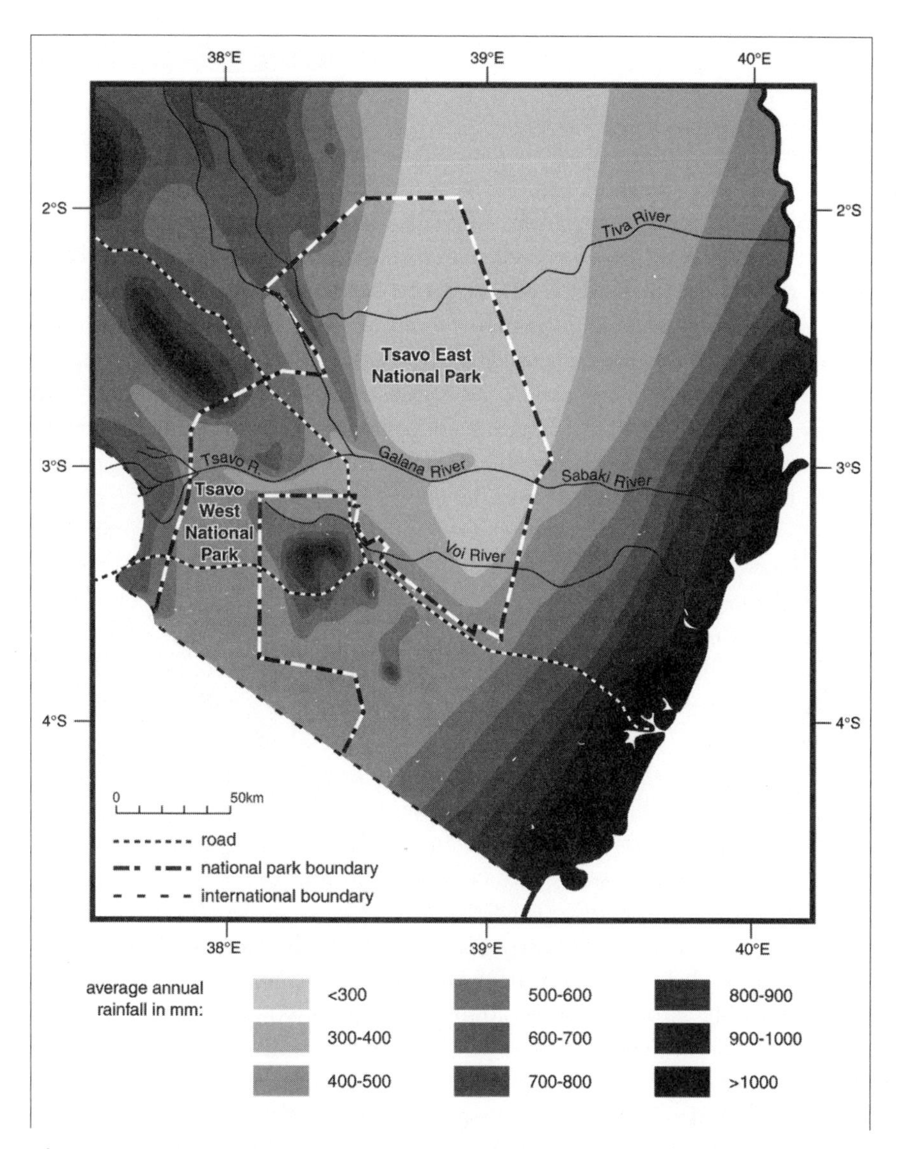

Figure 12. Precipitation contours (in millimeters) in southeastern Kenya, showing the extreme aridity of most of Tsavo East. Heavy precipitation is recorded in adjacent regions, including the Chyulu Hills to the NW, the Taita Hills between the two Tsavos, and the coastal lowlands to the SE. Maneless lions prevail wherever annual rainfall is less than 400 mm.

likewise inhabits an area receiving sixteen to twenty-four inches of rain annually. Because these more coastal areas lie at lower elevations than the park as a whole and likely experience even hotter temperatures, we have an exception to the general rule that manes should increase in size and profusion with elevation, one that underscores the importance of water (see Chapter 7).

These findings agree with the extensive variation in manes that Stevenson-Hamilton documented near what is now Kruger National Park, South Africa. "In the neighborhood of the Sabi River, for instance, in the north-east Transvaal, where the nights in winter are very cold, practically all the adult males lions are possessed of large bushy manes, sometimes yellow, but more often black or tawny.... The bush being everywhere thick and thorny, a forest life does not appear to have the deleterious influence on this hairy growth often attributed to it.... Some thirty miles north of the Sabi, where the country generally is fairly open in character, and where, consequent on the absence of any large body of water, the night temperature is much more equable, male lions are either maneless or possessed of light-coloured manes of small size..." (1912, 165–166). Although Stevenson-Hamilton was impressed by the difference between these areas in temperature, we would emphasize their differences in water availability.

The occurrence of well-maned lions in the Kalahari and Sahara deserts appears to refute this hypothesis. Various areas inhabited by well-maned lions receive as little rainfall as Tsavo or even less. Parts of Kalahari Gemsbok Park receive only about five and a half inches of precipitation annually, versus about ten inches falling in the center of Tsavo East (e.g., Sobo). And some of these areas may be even hotter, at least seasonally—the highest temperatures Eloff recorded in the Kalahari were 105.8° F (1973), while Voi has not had temperatures recorded in excess of 100.4° F. However, it is difficult to compare any two places in terms of the effects heat and aridity may have on the animals that live there, especially the roles played by environmental extremes. It is likely that Tsavo's extended dry season—predictably lasting four to eight months per year—is responsible for manelessness there. By restricting water availability to a few permanent sources, Tsavo's climate places a premium on heat tolerance and water efficiency, both of which would constitute potent selection pressures for the elimination of manes.

After the manuscript for this book was written, Peyton West and Craig Packer presented compelling evidence to show that the length and color of manes influence head loads (acquisition and dissipation of heat) in male lions (West and Packer 2002). Using infrared imaging equipment to monitor lion body temperatures, they determined that heat loads were substantially less at high elevations than in coastal regions, permitting manes to be correspondingly profuse. Curiously, however, they made no mention of water in discussing the costs of heat loads for lions. As a result, the causal mechanism they proposed for selection on manes is incomplete. Kays, Kasiki, and I believe that year-round access to water in Ngorongoro Crater may be as responsible for the long, dark manes of its lions as the crater's cool temperatures. We suggest that water deficits, especially during extended seasonal droughts, and not high temperatures per se, are responsible for the lack of manes in some populations of lions. Heavy manes exact water costs that limit the time periods lions can be active or their cruising radii around permanent water.

Developmental Hypotheses for Manelessness

Developmental or environmental hypotheses hold that Tsavo lions are potentially "normal lions" that would develop manes under different developmental or environmental circumstances. Unlike the evolutionary hypotheses, developmental hypotheses for manelessness would not entail major genetic shifts. The very modest genetic divergence of Tsavo lions uncovered by Dubach and colleagues is entirely consistent with these hypotheses.

The "Thorn-Scrub" Hypothesis

One hypothesis holds that dense thorn-scrub in certain habitats incessantly plucks the manes of lions inhabiting them. Anyone walking through East Africa's *nyika* or thorn scrub immediately appreciates the plausibility of this hypothesis as a cause of manelessness. In the more densely vegetated portions of Tsavo, it is difficult to walk for more than 100 yards without catching one's clothes on acacias and wait-a-bit bushes. So ubiquitous are they that scarcely a night could go by without a maned lion snagging some of his ornament on bushes along his path. In a very short time, a maned lion could lose all or most of his mane, and lions reared in such environments might never be expected to grow them. Swedish zoologist Einar Lönnberg called attention to this fact: "The Somali Lion is apparently a 'Bush-Lion' and this may account for the scantiness of its mane, such an ornament being decidedly not useful for an animal living in thick bush as it must get entangled in the thorns and twigs" (1912, 76).

Acclaimed photographer Martin Johnson also was aware of this phenomenon. In his book *Safari*, he wrote, "This picture shows two males and one female. They are fully grown, but the males have not yet developed big manes. It is doubtful if they ever will, as they are constantly getting the hairs full of thorns and burrs, which they comb out with their claws, pulling out the hairs at the same time" (1928, caption of plate opposite 272). John Hunter also noted that lions tore their hair out as they pushed through thick brush (1952), and Chris McBride routinely found tattered clumps of hair from the manes of his white lions of Timbavati clinging to thorn-scrub along their favorite trails (1977).

The congruent distributions of thorn-scrub habitats and scanty manes broadly supports this hypothesis. Lönnberg recognized that maneless lions inhabited the thorn-scrub desert inland from Mombasa and attributed this trait to the effects of thorns (1912). Dyer discussed variation in the appearance of lion manes and associated manelessness with lions inhabiting the "hot, thorny desert" (1973). And the Asiatic lion, well known for its sparse mane, is now restricted to the thorn-filled Gir Forest (Chellam 1996).

But at a finer geographic scale, this general association begins to break down. Frederick Selous confessed, "How to account for the variation in the length and colour of the mane in different individual lions I do not know. The theory, that it depends upon the density and thorniness of the jungles they inhabit, which pulls out and destroys their manes to a greater or lesser extent, I do not find tenable, as on the high open plateaus of the Matabele and Mashuna countries, where scarcely a thorn-bush is to be seen, lions of every variety as regards length and colour of mane are to be found, and the same variations also occur amongst those found in the neighborhood of Tati, where the country is for the most part covered with thick, thorny jungles." (1881, 258–259).

François Sommer also questioned the validity of thorn-bush as an explanation for manelessness. "It is often said that the lions of bushy and thorny districts have not such good manes as those that live in grasslands. This is probably true, although I have seen some very fine specimens with good manes in French colonial territory, particularly in the Chad.... " (1954, 132). And Astley Maberly, writing of South African lions, maintained similar beliefs: "The old idea that the thorns pull out the hairs so that a wild lion can never grow a really big mane is no longer valid, because lions with magnificent manes are often found in very dense and thorny country. Their manes, however, always look better groomed than those in captivity, because the thorny bushes certainly comb out the loose hairs—which so often become matted and unsightly in captive lions" (1963, 146).

Actually, the distributions of maned and maneless lions in Tsavo provide a compelling refutation for this hypothesis, in agreement with the positions espoused by Selous, Sommer, and Astley Maberly. Tsavo's vegetation has been subject to dramatic fluctuation over the last century, the area having been virtually choked with thorn-bush at the turn of the last century and denuded by drought and overbrowsing by elephants in the 1960s and 1970s. Nevertheless, throughout the last century, maneless lions have predominated in Tsavo East and extended into Tsavo West along an arid strip lining the Tsavo River. Areas to the west receive substantially more rainfall and support much denser shrubby and woody vegetation. Trees, shrubs, and bushes are generally larger, denser, and more abundant in Tsavo West than in Tsavo East. In fact, away from temporary watercourses, the southern portion of Tsavo East supports few bushes taller than 40 inches. Yet bushy Tsavo West supports moderately to well-maned lions, and most of Tsavo East is open but exclusively populated by males lacking manes.

The distributions of maned and maneless lions in Tsavo is in fact the reverse of what one would expect under this hypothesis from the distribution of thorn-bush. However, we cannot reject this hypothesis outright as an explanation for manelessness in Tsavo. During the long dry season of 1999, Roland Kays noted an abundance of dense mats of burrs produced by the plant *Pupalia lappacea* (Amaranthaceae) in many parts of the Tsavo East. *Pupalia* is a prostrate, scrambling, or erect herb native to Africa from the Cape north to East India. It is found in many areas where maneless or scantily maned lions have been reported, including the Gir Forest. Its burrs are incredibly sticky (Plate 4) and were noted adhering to the long hairs of lions, baboons, and jackals. Although the burrs are easily removed from short body hair, they become so entangled in longer hair that even dexterous baboons that devote substantial time to grooming each other could not remove them without uprooting hair. We suspect that lions could tear out their own manes when attempting to remove mats of burrs. Because we have not yet established the distribution and abundance of this burr plant throughout the greater Tsavo ecosystem, or in other areas within the lion's extensive range, we retain this modification of the thorn-scrub hypothesis as a possible alternative to explain manelessness there.

A test of the thorn-scrub hypothesis ought to be possible by examining histological cross-sections of the skin. Hairs grow from cell-lined cups in the skin called *follicles*, and the follicles of different hair types exhibit different morphologies, whether or not they currently contain hairs. Therefore, studying sections of skin from maned and maneless lions should determine whether the density and structure of mane-hair follicles differ in the two groups. If so, there would be microanatomical evidence that maneless lions differ from maned lions in ways other than the vegetation densities of their home ranges. Differences in follicle density or structure could be based on either genetic or developmental causes. The latter might include chronically depressed or elevated testosterone levels (see below). Kays, Kasiki, and I are currently pursuing these questions in collaboration with British endocrinologist Julie Thornton of Bradford University in the United Kingdom.

The "Down-and-Out" Hypothesis

Another "nurture" hypothesis holds that manelessness may result from injury, poor condition, or old age (see, e.g., Hunter 1999). Such conditions may cause altered levels of testosterone, the male hormone responsible for mane production, and possibly lead to deteriorated condition or loss of the mane. Poor nutrition and chronic domination by other lions might cause nomadic lions to exhibit low levels of testosterone and to reflect this in their scanty or ragged manes. Obviously, this hypothesis depends on the mane's serving as a signal of general condition and is compatible with the "truth in advertising" model of sexual selection. Males with scanty manes would be unmistakably branded "losers" in the scramble to leave genetic descendants. Unfortunately, there is little direct evidence on hormone levels, mane conditions, and associated behavior patterns in free-ranging lions; but existing data from other regions tend to support the hypothesis. For example, West and Packer recently showed that in maned lion populations, poor manes were commonly associated with injuries and poor nutrition (2002).

Critical data on hormone levels and fitness in Tsavo are lacking, but at face value, the "down-and-out" hypothesis appears to be false. While Tsavo's lions are probably not significantly larger than other East African lions, they are certainly no smaller. Large size in male lions is achieved by an extended growth period that continues after the onset of puberty and that carries them from sexual to social maturation. This period takes them beyond the largest body sizes attained by females. Testosterone produced by the maturing testes is crucial both to the development of this size dimorphism and to the production of manes. Well-maned lions that are castrated promptly lose their manes (Guggisberg 1961; Pocock 1931). Tsavo lions have the large body size characteristic of sexually and socially mature lions elsewhere but lack the mane that is routinely associated with it. In addition, the Tsavo population is surrounded by populations of maned lions whose boundaries are dynamically maintained by dispersal, competition, and natural and sexual selection. Were the entire Tsavo East population composed of losers, as this hypothesis suggests, manelessness would quickly disappear, swept away by competition with fitter and more successful maned populations.

Janine Brown and colleagues measured hormone levels in free-ranging male lions to assess their reproductive function and cycles. They compared prepubertal, young adult (3–4.5 years) and adult (6–10 years) males from the Serengeti Plains and Ngorongoro Crater, Tanzania. In the Serengeti, age affects semen quality. Young adult males produced fewer normal spermatozoa and sperm of lower mobility than adult males. Both the adult and young adult samples examined were collected from lions resident in prides, eliminating social context as a covariate; both age groups are part of the breeding population. The results are noteworthy, because in most other species, maximal testosterone secretions peak at or shortly following puberty. But in polygynous animals such as cattle, ejaculate volume, sperm concentration and motility, and percentage of normal sperm all increase following a bull's sexual maturation. This type of age variation certainly contributes to variation in fertility and probably influences the reproductive success of breeding male lions (Brown, Bush & Packer 1991).

The chain is actually a bit more complex. The hypothalamus secretes a small peptide, gonadotropin-releasing hormone (GnRH). This hormone regulates the release of the luteinizing hormone (LH) from the anterior pituitary gland, which stimulates the gonad. LH in turn stimulates testosterone production by the testes.* Although GnRH and LH levels in young adult lions are high—comparable to their levels in adults—the testes do not reach peak testosterone production for 1–2 years. Enhanced testosterone production was achieved via a higher concentration of LH receptors in the testes rather than by elevated serum concentrations of the hormone (Brown, Bush & Packer 1991).

Differences in testosterone levels have not yet been tied to lion behavior. Peak testosterone activity occurs in adult, not young adult, lions, and naturalists have noted age-related differences in aggression or belligerence. Hunter described an incident in which a male lion, surrounded by spearmen, preferred to flee rather than fight. "I have often observed that the old lions with the finest manes are more reluctant to give battle than young males or females. The same is true of elephants.... I suppose they learn discretion with the years" (1952, 105). Astley Maberly corroborated this opinion. He said, "In most parts of Africa the Natives distinguish two types of lion: A large, heavy-maned beast, quite unmarked; and a smaller type with only a scruffy mane, and usually more or less 'spotted' on flanks and limbs

*http://www.merck.com/pubs/mmanual/section18/chapter234/234a.htm

(particularly in the case of lionesses). They aver that the smaller, spotted type is the most dangerous and aggressive.... Most game rangers agree that such distinctions are merely a matter of age, the younger animals still displaying the remains of infantile markings, undeveloped manes, and inclined to be more reckless, inquisitive, and sometimes more aggressively inclined than their more experienced elders" (1963, 147).

Perhaps Hunter had it right and these behavioral differences are simply a matter of experience and learning. But it is reasonable to suppose that behavior is also affected by hormone levels. Testosterone levels affect territoriality and aggression toward members of the same species (Millar et al. 1987; Rachlow, Berkeley & Berger 1998; Woodroffe, MacDonald & Cheeseman 1997). However, elevated aggression levels (Hubert 1990; Lumia, Thorner & Mcginnis 1994) need not be a general response toward other living creatures. Here again, aggression is a behavioral concept best applied to interactions between members of a given species (Lorenz 1966).

The "Über-male" Hypothesis

It is even possible that manelessness could result from too much testosterone rather than not enough of it. Although testosterone affects hair growth, its effects on hair follicles are indirect and mediated by the follicle's dermal papilla. In the early 1990s, Julie Thornton of Bradford University and colleagues compared androgen-sensitive beard follicles with non-balding scalp follicles that are less androgen-dependent (Thornton 1993). They found that androgen-sensitive cells contain higher amounts of 5-alpha-reductase, an enzyme that transforms circulating serum testosterone into a highly active form of testosterone called dihydrotestosterone (DHT). DHT affects many aspects of male behavior, from aggression to sex, and is the primary agent of male pattern baldness. Over time, DHT causes a hair follicle to degrade. Although some follicles die, most shrink in size, producing progressively finer unpigmented hairs. Eventually, these hairs become too weak to survive daily abrasion. However, only those cells producing the reductase enzyme undergo this transformation, accounting for the very characteristic pattern of this baldness (horseshoe-shaped and confined to the crown). Both the testosterone and the reductase are essential for this response: Eunuchs never go bald, regardless of their genetic backgrounds, but castrated men who are given testosterone can develop bald spots.*

*A much fuller treatment of this complex subject can be found at http://www.health-library.com/hairloss.html.

Is it possible that the Tsavo lions have a form of male-pattern baldness that affects only their mane hairs, leaving the remainder of their physiology and behavior intact? Balding men don't have abnormally high levels of circulating testosterone but rather possess above-average amounts of DHT in the affected scalp follicles. We have begun collecting skin biopsies taken from the neck areas of maned and maneless male lions to assay reductase concentrations and will compare these with serum testosterone levels. These samples are collected via fieldwork that is jointly funded by the National Geographic Society and Earthwatch Institute (Plate 2).

Higher-than-average testosterone levels in Tsavo's lions are possible and could offer a unified mechanism for explaining their morphology, ecology, and behavior. Relatively high levels of testosterone may have evolved as a spacing mechanism, to limit lion density to levels commensurate with prey abundance. Increased aggression by male lions toward intruders in their territories may raise the stakes of male-male conflict and permit solitary males to fend off groups of challengers. Elevated levels of testosterone in Tsavo males could be associated with the absence of male coalitions in our surveys; male-pattern baldness of the mane might have been an incidental by-product of selection's acting directly upon hormone levels (Ebling 1987), one that proved advantageous in views of the region's heat and drought.

Conversely, chronically depressed levels of testosterone could give rise to adult males that retained juvenile traits, including aggressive responses toward humans (accounting for the legendary aggressiveness of Tsavo's lions) and an essentially maneless condition. Only data can construct reliable linkages among these alternatives. We have recently enlisted Nadja Wielebnowski, a reproductive endocrinologist at Chicago's Brookfield Zoo, to help us evaluate androgens and adrenal hormones in Tsavo's lions. Wielebnowski, who has extensive experience with hormones in other cats, including snow leopards and cheetahs, could assay hormones in Tsavo lions from blood and scat samples. The surveys that Kays, Kasiki, and I have initiated are collecting samples that should permit Thornton and Wielebnowski to distinguish between these alternatives.

Manes as a Target for Sexual Selection

Manes constitute a model system for evaluating models of sexual selection and specific hypotheses for trait evolution. However, far more information is needed to tie variation in manes to physiology, behavior, and ecology and ultimately to male reproductive success. Only then will it be possible to determine whether manes represent an honest advertisement of genetic quality, a self-imposed handicap to demonstrate hardiness, or the product of female whimsy. The recent report by West and Packer was an important step in this direction (2002), but sexually selected features of other organisms remain better understood.

For example, numerous studies have explored the role of deer antlers in mating biology and their correlations with other life functions. Like the manes of lions, antlers in male deer are used in displays to both sexes and in combat with other males. Like manes, antlers develop under the influence of testosterone. During different phases of the antler cycle, the testes may triple or quadruple in size to produce the necessary hormone triggers. Like manes, antlers may serve as targets of sexual selection either by female choice or by male-male contest. As with mane size and color, there is great variability in the size and branching pattern of antlers. Much of the variation in both antlers and manes is a function of age, health, and environmental factors.

Antler size and number of tines in red deer steadily increases with age up to five to seven years, grows more slowly to age twelve, and then declines in senescent individuals thirteen to sixteen years old. This variation parallels serum testosterone levels—female white-tailed deer experimentally treated with testosterone grow antlers (Sauer 1984). Antler size also varies with social status and with nutritional condition, in ways that suggest regulation by testosterone. Antlers are expensive to grow and to maintain, as they are shed and must be grown anew each year. During peak growing season, males have even greater energetic requirements than pregnant or lactating females. Finally, antler size is correlated with body weight (a measure of condition) and frequency of roaring displays during the rut. This trio of variables is inextricably linked to male reproductive success in red deer (Clutton-Brock, Guinness & Albon 1982). The close association of large antler size, large body size, and high bellowing frequency means that antlers reflect overall genetic quality in ways that should benefit both male and female offspring (Kodric-Brown & Brown 1984). Many view it as a superb example of "truth in advertising."

Mane quality in lions also appears correlated with a suite of characters associated with greater fitness, including nutritional intake, fighting ability, and female mating preferences (West & Packer 2002). On the other hand, lions in the Ngorongoro Crater are famous for their long, dark manes (Sommer 1954), but their testosterone levels are no higher and their semen quality significantly lower (Munson, Brown & Wildt 1996) than among males on the Serengeti Plains. And lions in Tsavo are practically maneless yet are surrounded by maned populations. Given the female preferences documented by West and Packer, why don't Tsavo females preferentially breed with maned males? Why haven't maned males invaded Tsavo and run out the genetically or physiologically inferior residents? Perhaps manes are affected by a wider suite of factors than antlers, but antler development is also influenced by light and latitude, various hormone levels, genetics, and nutrition (Goss 1983). Obviously, far more information is needed to evaluate components of mane variation.

Chapter 9
Conservation and Tsavo National Parks

Hector, you must be mad to talk to me of a pact. Lions do not come to terms with men, nor does the wolf see eye to eye with the lamb – they are enemies to the end. Achilles, on the gates of Priam (Homer, *The Iliad*, Book XXII, 404).

Throughout history, humans have waged a relentless war on lions. Over the last two centuries, the geographic range of lions has shrunk dramatically, as they were extirpated from many places they formerly occupied. Expanding human populations and resource needs are partly responsible, for they have reduced living space, thinned natural resources, and eroded resource quality for lions and many other organisms. However, apex predators like lions have suffered disproportionately. Besides losing habitat and prey to preemptive use by humans, lions became targets of direct persecution by preying on livestock and attacking people.

Conserving biological diversity is surely the most pressing environmental challenge of the twenty-first century. The health of biodiversity is both a measure of our stewardship and a gauge of the sustainability of our lifestyles—indeed, our life-support systems. Most wildlife species share the lion's uncertain prospects in our human-dominated world. In this concluding chapter, I consider biological conservation and its future, with special reference to Tsavo and its lions.

The Conservation Imperative

Even during the railway's construction, colonial administrators began to recognize the need for conserving Africa's extraordinary fauna. Railroads and highways opened the interior of Africa to foreigners, and a wave of colonization ensued—people from Europe and America and Asia attracted by the wildness and beauty of primeval Africa. Invariably, soon afterwards, the interior was sapped by unsustainable human demands for its natural resources. As Peter Beard noted in *The End of the Game,* "The deeper [white men] went into Africa, the faster life poured out of it, off the plains and out of the bush and into the cities, vanishing in acres of trophies and hides and carcasses" (1988, 112).

Although Beard's passionate language suggests hyperbole, one has only to scan "Africana" literature to appreciate at first hand the bloodthirstiness of Africa's early sportsmen. Between 1877 and 1880, a single hunter—Frederick Selous in *A Hunter's Wanderings in Africa*—recorded killing 548 game animals on his forays, including 100 buffalo, 20 elephants, 48 zebras, 42 hartebeests, 39 eland, 33 sable antelope, and multiple individuals of 26 other species (1881). In the appendix of his book on the man-eaters, Col. Patterson prescribed a virtual recipe for faunal endangerment, listing the supplies needed by a sportsman for a three-month shoot in the East African bush. His list included three guns (.450 express and .303 sporting rifles and a 12-gauge shotgun), together with over a thousand rounds of ammunition: 250 rounds of .450 (for very large game), 300 rounds of .303, and 500 shells filled with #6 and #8 shot for small game and birds (1907, 325). Patterson and his friends even shot game from moving trains! It wasn't long before impressive early bags gave way to widespread concern for wildlife and governmental control.

Noel Simon has written a superb account of the early conservation movement in Kenya, entitled *Between the Sunlight and the Thunder: The Wild Life of Kenya.* His book offers a detailed history of early efforts to conserve wildlife and the individuals responsible for them. Certainly, in conservation terms, Kenya is today the beneficiary of the colonial stewardship of the British. From their first forays in East Africa, the British acknowledged the need for conserving wildlife. That conservation ethic in turn gave rise to the country's system of parks and reserves. Under "The African Order in Council, 1889," the Queen's Regulations made breaching the game laws a punishable offense in the British East African Protectorate. An 1897 law prohibited shooting of game within a twenty-five-mile radius of a Government Station, except as permitted by district officers. On 11 August 1899, the Foreign Office proclaimed new game regulations, which established the "Ukamba Game Reserve." This was originally decreed to cover "the whole of the Kenia District of the Province of Ukamba, except the area within ten miles around the Government Station at Kikuyu...."

(1962, 116). In April 1906, this was renamed the Southern Game Reserve and reconstituted to cover 10,695 square miles. The reserve bordered the Tsavo River on the east, the Uganda Railway on the north nearly to Nairobi, the Mau Escarpment on the west, and the Tanzanian border on the south. It afforded protection to a host of wildlife species. It even drew the favorable notice of President Teddy Roosevelt: "The English Government has made a large game reserve of much of the region on the way to Nairobi, stretching far to the south, and one mile to the north, of the track. The reserve swarms with game; it would be of little value except as a reserve; and the attraction it now offers to travellers renders it an asset of real consequence to the whole colony" (1909, 13).

Blayney Percival was the protectorate's first game warden. His reports confirmed that the Southern Game Reserve was full of game but also that it was under assault, even a century ago. The most serious challenge came from the Chagga people of northern Tanzania, who had riddled the base of Mount Kilimanjaro with game pits. The Wandorobo, one of Kenya's two hunter-gatherer tribes, also hunted in the Southern Reserve, but they had little impact on game populations because their numbers were small and their harvest methods were sustainable. The populous Wakamba offered a conspicuous contrast of intensifying impact. Historically, the Kamba people had been excluded from much of the Southern Reserve by the bellicose Maasai, who ruled the open plains in what is now southern Kenya and northern Tanzania. However, with the collapse of Maasai dominion in the wake of the pleuropneumonia and rinderpest epidemics of the 1890s, the Wakamba began to exploit the Southern Reserve. Their subsistence hunting intensified as recurring epidemics and droughts decimated their own herds. They hunted in large parties and killed game in droves—any incentive for sustainable methods disappeared as they began to exploit distant and uncontrolled hunting grounds (Simon 1962). Several species, including Grant's gazelle, began to show signs of overexploitation. Of course, buffalo, kudu and eland populations were all dangerously low, having been hammered directly by the epidemics themselves.

Although the traditional technologies of native hunters did not decisively alter the conservation balance, they were far more sophisticated than arrows or spears. Pitfalls dug along game trails captured animals indiscriminately, including even elephants. Drop spears were used to prevent animals from escaping them. Thorn fences, some more than a mile long, were used to channel game. Bowmen hidden in and behind the fence would shoot poisoned arrows into animals driven by beaters or wildfires. Recreational hunting by European and American sportsmen added to this relentless impact.

By 1906, the Earl of Elgin restricted the number of sportsmen's licenses issued annually to five hundred, and placed progressive restrictions on bag limits. Whereas only their ammunition supplies and coterie of porters had limited earlier hunters, twentieth-century hunters began to face quotas designed to control the harvest of vulnerable species.

During the First World War, the Southern Reserve became contested territory as troops massed on either side of the border between British Kenya and German-controlled Tanganyika. Subsistence hunting became a major factor in game declines: Percival estimated that at least 40,000 game animals were harvested for meat over a two-year period. In addition, black rhinos were shot on sight, not for meat but for self-protection: Rhino trails offered the only unobstructed passageways through the region's endless wait-a-bits and acacia, and people preferred walking on them. Hundreds of rhinos—more than live today in all of Kenya—were shot and simply left to rot. Giraffes were also shot as vermin because they often became entangled in telephone and telegraph lines, disrupting communications (Simon 1962). The exploitation of wildlife and native lands did not end with the war, and the Southern Reserve suffered additional boundary adjustments and changes of status during the postwar period. Declining numbers of wildlife, impaired water supplies, and escalating herds of livestock reduced native vegetation in many areas to near-desert conditions. Additional measures were obviously needed to conserve the nation's wildlife in the face of national development.

The Creation of the Tsavo National Parks

Kenya's national park system came into existence at the close of World War II. Captain Archie Richie, Game Warden of Kenya, was instrumental in establishing it. Richie was appalled that political expedience could redraw at will the boundaries of game reserves, thereby exposing the animals within them to exploitation. Accordingly, he advocated establishing sanctuaries with immutable borders, what the world was only beginning to recognize as national parks.* Initially, Kenya's planners envisioned three national parks: the Northern Reserve, the Southern Reserve, and a section of the Aberdares Mountains (Simon 1962).

*In *The Man-eaters of Tsavo*, Col. Patterson recounted a hunt in which two lions were left dead overnight where they lay—the next morning, two other lions were found feeding on one of these carcasses. Observers have also witnessed natural and uninterrupted instances of cannibalism—in one, a male killed a female in a squabble over a kill, then consumed much of her in preference to the prey (Fitzsimons 1919).

According to Simon, the idea to include Tsavo as a part of the park system originated with Assistant Warden C. G. MacArthur. In 1939, MacArthur suggested incorporating as a park the entire region between the Wakamba and Maasai reserves and between the coastal strip and the Tanzanian border. His plan was initially well received, because the enormous region had only a handful of legal inhabitants and few competing claims. However, the area slated for protection was successively reduced as different interest groups excised sectors with the potential for economic development. Worse, the cuts were made in a fashion that impaired the park's effectiveness for conservation. As Noel Simon lamented, "The boundaries, having been pruned and lopped to avoid including areas which might at some future date prove useful for an alternative purpose, resulted in an ill-gotten and strangely misshapen piece of land, as a glance at the map will show. A similar-sized region, properly shaped and more compact, would have been infinitely more valuable from a conservation standpoint.... Eight thousand square miles is an immense stretch of country, and space is without question an extremely important factor in its own right when designing national parks. But it is entirely fallacious to assume that space, or, for that matter, remoteness, are sufficient in themselves to provide adequate sanctuary. This might not be so bad if it were not for the fact that the Tsavo park is the *only* major national park in the whole of Kenya, and the evidence available makes it abundantly clear that it became a park only because it was so poor and worthless" (1962, 127).

Daphne Sheldrick, the celebrated wife of Tsavo's first warden, David Sheldrick, offered a comparable appraisal. "[Tsavo] was set aside as a national park not because it harbored a wealth of wild animals, which it certainly didn't, nor because of any spectacular scenic advantage, but simply because it was the only empty piece of land that could be spared, being unsuitable for other purposes due to aridity, a lack of permanent water sources, and the presence of trypanosomiasis [sleeping sickness, a disease caused by a parasitic flagellate protozoan]. The only people who penetrated that largely unknown chunk of Africa then were small bands of poachers from the Waliangulu and Wakamba tribes who made a living by hunting elephant and rhino, selling the trophies to illicit dealers in Mombasa" (2000, 28). Tsavo was not a reserve set up for the benefit of game species or for sports hunters—Tsavo was what Kenya was willing to spare for wildlife at that time in the nation's development.

Certainly, biodiversity scientists would have configured the park quite differently. A population center in the midst of the greater Tsavo ecosystem—the town of Voi and densely settled Taita Hills*—dramatically reduces the sanctuary's effectiveness. Such inclusions mean that the reserve has both internal and external "edges," increasing its susceptibility to external forces, especially those created by humans. But the exclusions from the park are significant—in and of themselves—quite aside from the risk of human incursion. The Taita Hills, Sagalla, and Kasigau are the northern outposts of an extensive system of ancient mountains stretching across the entire length of Tanzania. The Eastern Arc Mountains and coastal forests of Kenya and Tanzania support faunas and floras of very high distinction and diversity; they even rank among the world's twenty-five "biodiversity hotspots" (Myers et al. 2000). *Purposefully excluding* these areas from the Tsavo Reserve was not a decision based on their perceived conservation value. Rather, because they intercept twice as much rainfall as surrounding areas, these areas had already been densely settled by the time the park was being created.

Despite its shortcomings, Tsavo is a magnificent sanctuary by any definition. In terms of area, all of Kenya's other national parks can be encompassed within the boundaries of Tsavo East and Tsavo West National Parks. Their size and wildness have invited managers to preserve other areas in the region, annexing them into a regional conservation strategy. Currently, Tsavo East and Tsavo West National Parks are bordered and buffered by the Chyulu Hills National Park to the northwest, South Kitui National Reserve to the northeast, and Tanzania's Mkomazi Game Reserve to the southwest. A private consortium of ranches being managed for conservation, the Tsavo-Kasigau Wildlife Corridor, is developing to bridge the 772 square miles between the southern arms of Tsavo East and Tsavo West.† Collectively, the greater Tsavo ecosystem represents one of the very largest protected areas in all of East Africa. Moreover, it lies astride Kenya's busiest highway and is home to the Kenya Wildlife Service's paramilitary school for training rangers at Manyani. This location means that the parks support not only the largest populations of many wildlife species, including elephant, rhino, and probably lion, but also possibly the most secure.

Large wilderness area, varied landscapes, and high security mean that Tsavo can offer haven to some of Africa's most delicate conservation projects. Both Tsavo East and Tsavo West support freeranging populations of

*The town of Voi had 24,532 registered voters in the 1997 general election, according to the *East African Standard* (http://www.eastandard.net/elections2002/hotspots/coast/taita_taveta/voi.htm); perhaps as many as one hundred thousand people reside within the Greater Tsavo ecosystem.

†http://www.savannahcamps.com/tdc/wildlifecorridor.html

black rhinos and have rhino breeding programs. The parks support large herds of elephant that are recovering nicely from decades of drought and poaching; 8,068 elephants were documented in a January 1999 census of Tsavo, a 66% increase over their numbers a decade earlier (Kahumbu et al. 1999). And Tsavo East is home to the sole protected population of hirola (Hunter's hartebeest), arguably the world's rarest antelope (Andanje & Ottochile 1999). Although stewardship of these programs requires constant vigilance, their presence in Tsavo attests to its enormous value for conservation.

Area is one of the most essential features for natural conservation, ensuring large, healthy populations of protected species. Large populations are less susceptible to the sorts of environmental, demographic, or genetic accidents that can cause smaller ones to flounder. Critically, large areas support the full gamut of ecosystem processes—including nutrient cycling and different stages in ecological successions—that are involved in ecosystem maintenance. This is a critically important issue for conserving African savannas. The grazing activities of antelope and buffalo are known to thin Africa's grasslands and encourage the proliferation of herbaceous and woody plants, which initiate the development of woodland formations. In turn, the foraging activities of elephants, rhinos, and other browsers prune the woodlands, fostering grasslands, and maintaining a healthy equilibrium (McNaughton & Georgiadis 1986). Because of the large home ranges of the mammals that perform these ecosystem services and the need for maintaining healthy populations of them, huge preserves are needed to maintain savanna environments through time. Savannas are a heterogeneous mosaic of grassland and woodland patches, maintained as an admixture by herbivores, fire, and variation in climate and soils (Belsky 1992).

In *The Song of the Dodo*, author David Quammen considers the numerous ways in which a park's area, isolation, and other attributes affect its ability to safeguard biological species (1996). Sadly, many national parks in Africa and elsewhere are too small to protect their inhabitants over the long term. Already, species have disappeared from North America's parks during their short history (Newmark 1995), and some of Africa's premier national parks are far too small to offer safe haven over the long term. For example, Kenya's Amboseli National Park, home to some of the world's best-studied elephant populations, made famous by Cynthia Moss, David Western, and Joyce Poole, measures only about 150 square miles. Nairobi National Park, Kenya's oldest national park, offering unparalleled opportunities to view black rhinos in the wild, is even smaller (117 km^2 [45 square miles]) and is now nearly completely surrounded by urban and suburban sprawl.

No Park Is an Island

Although parks frequently appear to be "islands" of natural vegetation in a "sea" of human-dominated lands, this analogy is only approximate. True, the lands surrounding national parks are often less favorable for park inhabitants, but they aren't uninhabitable. A host of interactions take place between park denizens and inhabitants of the surrounding matrix. One of these is indisputably beneficial; most are distinctly perilous.

On the positive side, parks support only a fraction of Kenya's flora and fauna. Actually, most of Kenya's wildlife is thought to reside on lands outside its national park system (Western 1997). Continuity of park and matrix populations enlarges the effective population size of many species, and this reduces inbreeding, loss of variation through genetic drift, and chances of local extinction. Where peripheral populations sporadically disappear, they are quickly reestablished through natural recolonization.

But wildlife populations are not the sole residents of the matrix between parks. Increasingly, these intervening areas are crawling with humans, their domesticated dependents, and introduced species acting as competitors, pests and pathogens. All reduce the effectiveness of parks by encroaching on protected areas in various ways, reducing their effective area and the population sizes of native species dependent on them.

Human encroachments are obvious, because we are attuned to human activities and can easily perceive them. Park-goers in Africa frequently find herdsmen with cattle or goats on the margins of parks and reserves or well within their borders. Such trespass may be incidental or episodic. To maintain good community relations and ensure local cooperation, parks sometimes grant local residents access to permanent water within the park during particularly severe droughts. On the other hand, trespassing the park's borders may be habitual, and grazing or browsing within the park may be an essential feature of herd management. Such activities reduce the carrying capacity of the environment for herbivorous species of wildlife. More significantly, grazing and watering stock within parks may foster the transmission of disease and parasites and provoke human-wildlife conflicts with stock raiders.

A more insidious form of human encroachment is perceptible along the Nairobi-Mombasa Highway between Ndii and Mackinnon Road: Gunnysacks of charcoal line the roadway, on sale to travelers in both directions. Charcoal is a lightweight, smokeless, and almost ash-free cooking fuel that is produced by burning selected types of wood, especially species of *Acacia*, in the absence of oxygen. Although commercial production and sale of charcoal are strictly controlled, any rural person with an axe, ambition, and access to woodlands can generate this cash crop. Current extraction rates are largely uncontrolled and almost certainly unsustainable. Sadly, charcoal producers prefer the same species of trees that are coveted by elephants and giraffes, so this human use competes directly for browse with Africa's largest herbivores. Even without a booming charcoal industry, people questioned whether Tsavo's woodlands could sustain its sizable elephant population (Laws 1974). Excessive harvest of slow-growing savanna trees for cooking fuel is also a problem in the areas bordering the Masai Mara Game Reserve in southwestern Kenya.

Deforestation and the extinctions that result from habitat loss are a fundamental issue for conserving Kenya's forest-dependent species. This is the most serious risk to the faunas and floras of Kakamega and Arabuko-Sokoke forests, Mount Kenya, the Mau Escarpment, and in the Taita Hills (Brooks, Pimm & Oyugi 1999; Lens et al. 1999). But the biggest threat to Kenya's savanna wildlife now comes not from habitat conversion but from "bushmeat" or subsistence hunting, one of the greatest and most rapidly growing threats to tropical mammal and bird species worldwide (Plate 9). Using firearms, snares, pitfalls, and other capture devices, rural people and market hunters harvest wildlife at unsustainable levels. Currently, in places like the Congo Basin, as much as a third of the standing biomass of mammal species may be harvested annually, a rate that far exceeds production. Organizations like the Bushmeat Crisis Task Force, an international consortium of zoos, NGOs and museums, have rallied to stem this rising tide of wildlife peril (see www.bushmeat.org).

According to Youth for Conservation, a Kenya-based conservation organization that focuses on the bushmeat issue,* poaching takes place in all Kenyan national parks. Typically it takes the form of snaring, as quiet and impersonal as it is insidious and effective. In February 2001, Youth for Conservation learned of active snares in both Tsavo parks and sales of bushmeat in the towns along the Nairobi-Mombasa Highway. Accordingly, they organized an expedition of youths to locate and remove snares. Over the course of a week, they uncovered 142 snares in an area approximately nine square miles in size. Eighteen of the snares were made of thick wire, intended to trap eland, buffalo, or giraffe, while the vast majority were made from smaller-gauge wire and intended for anything from zebra to dik-dik. Based on Youth for Conservation's findings, a snare catches an animal

*Josphat Ngonyo, Director, Youth for Conservation, P.O. Box 27689, Nyayo Stadium 00506, Nairobi, Kenya (http://www.youthforconservation.org/)

every twenty to thirty days. In some areas, such as the Taita Hills Game Sanctuary adjoining Tsavo West, poachers use wooden fences to channel game directly into their snares, sometimes taking animals as large as giraffes. Most snaring uncovered in Tsavo by Youth for Conservation could be attributed to the communities living immediately adjacent to the parks, including institutions such as Manyani Prison and the railway stations.

Although poachers do not target lions, snares sometimes capture cats. If they survive and manage to pull themselves free, lions may suffer a protracted death by starvation. Their incapacitation may lead them to become marauders or man-eaters. Honorary Warden Tony Seth-Smith showed me pictures of a lion he shot on Galana Ranch that had a snare cinched tight around the skin and muscle of its neck, a chronic cause of debilitation that virtually guaranteed that it would become a problem lion. One week before the Tsavo desnaring operation, a lion died in a snare near Itani in Tsavo West, and Youth for Conservation rescued a snared lion in their Masai Mara sweeps.

Repeated sweeps are necessary to discourage poaching in Tsavo and elsewhere—snaring is too effective and too profitable to curtail with one well-publicized survey, whatever the consequences. Monitoring and control must be predictable and effectively continuous. Efforts to intercept meat sold along the highway and prosecute violators must be stepped up. Educational efforts are also necessary to sensitize and educate residents on wildlife conservation. Most regard wildlife as "God-given meat" and its sole benefit as being the meat it provides. Few recognize that crop damage by animals is episodic, paling alongside the damage wrought by insects, drought, or blights. Fewer appreciate the indirect benefits wildlife brings to their communities through eco-tourism. The Tsavo region benefits from the educational efforts of the Tsavo Education Centre in Voi, the new visitors center in Mtito Andei, and the Taita Discovery Centre and Wildlife Works on Rukinga Ranch. The community education programs of the David Sheldrick Wildlife Trust helps to extend their impact. All strive to incorporate communities in wildlife conservation efforts.

Animals Are Not the Only Victims

The porous boundaries of parks are not solely a problem for the animals that reside in them. Just as humans and their stock can invade national parks, so does wildlife move from protected areas onto private lands. And when animals ruin crops, destroy livestock, or kill villagers during these sojourns, the resulting conflict between animals and humans threatens local economies and regional development as well as conservation plans.

University of Utah ecologist Bill Newmark and colleagues surveyed people living adjacent to some of Tanzania's premier wildlife areas, including Arusha, Kilimanjaro, Tarangire, Lake Manyara, Ruaha, and Serengeti—Ngorongoro Conservation Area. Of those questioned, 71 percent reported problems with wildlife, and the frequency of such problems depended on the density of human settlements around parks. Most people (86 percent) reported crop damage from game, while 10 percent reported that wildlife had killed livestock and poultry (Newmark et al. 1994).

Throughout the development of Kenya's parks and reserves, its government has been responsible for managing wildlife conflicts. "Game control" represents those measures needed to ensure public safety and economic development in the midst of wildlife. Game control was an essential element of conservation for Capt. Richie, Kenya's Game Warden from 1923 to 1949: "No human community will tolerate in its vicinity the existence of—much less subscribe to the protection of—species that are a perpetual source of danger or depredation; ... active intervention must always be ready and at hand" (Hunter 1952, xi–xii).

In Kenya's early days, wildlife seemed infinite; animal control was enthusiastic and unabashed. Its grim necessity is given expression in this quotation from Theodore Roosevelt's *African Game Trails:* "... during the last three or four years, in German and British East Africa and Uganda, over fifty white men have been killed or mauled by lions, buffaloes, elephants, and rhinos; and the lions have much the largest list of victims to their credit. In Nairobi church-yard I was shown the graves of seven men who had been killed by lions, and of one who had been killed by a rhino" (1909, 74). In such times, animal control was usually lethal. Control officer Alan Black is said to have decorated his hat with the tail tufts of fourteen man-eating lions he had killed (Hunter 1952).

As more and more of Kenya has become settled, nonlethal means of control have received increasing emphasis. Game control officers now routinely explore several options when dealing with problem wildlife: training animals to deter them from offensive behaviors, transporting animals from areas where they conflict with people, culling individual perpetrators, and locally extirpating problem populations. The last three are in use in most areas of East Africa, but their effectiveness varies with context, scale, and the particular species involved. Their efficacy is apt to depend on the individual animals to which they are applied. For example, translocation may be effective for the occasional stock raider, but culling is warranted for a habitual marauder (Stander 1990).

The actions of animal control officers are often misunderstood by persons who do not understand the extent of animal depredations or of the capacity of natural populations to increase. Even members of the conservation community sometimes regard such persons as misguided and out of touch with the times. Consider the anguish of Alistair Graham and Peter Beard when they were criticized for harvesting individual crocodiles to assemble a conservation plan for the Lake Turkana population: "To many sentimentalists the preservation of animals is a romantic cause.... Theirs is a world of heroes and villains, no less among the animals than among the men. Heroes do that which they dream of, and with a flourish. Wildlife biologists equipped with firearms and mathematical probability more closely resemble villains, the bandits of sentiment and fancy. Population biologists cannot afford themselves the luxury of getting involved in the lives of individual animals when it is the destiny of the whole species that concerns them" (Graham & Beard 1973, 12). In the same way, animal control often calls for eliminating noxious individuals in the interests of sustaining the coexistence of people and wildlife.

Tsavo's populous center exposes a large number of people to the diverse faunas the park system contains. As in Tanzania, elephants are responsible for the most complaints and the greatest economic damage, according to Tsavo East's scientific officer, Dr. Samuel Kasiki. At night, elephants leave the Tsavo parks to raid orchards and fields and disrupt aqueducts; occasionally, they even kill livestock. To safeguard the town of Voi, KWS has erected electrified fences, which force elephants leaving the park to circumvent the most populous regions, and selectively employs flash guns to condition avoidance behaviors.

In the Tsavo region, human conflicts with lions rank second only to those with elephants. Most involve lion attacks on cattle, sheep, or goats, although occasionally herdsmen or (more rarely) villagers fall victim. Some of the attacks are by park lions, which may trail herds that illegally grazed inside the park during the daytime back to their *bomas* at night. However, most involve lions that live outside the park boundaries, on the ranchlands that border the national park. Each year, such attacks claim the lives of hundreds of livestock and a handful of people.

Until his death in 2001, William Mukabane was an animal control officer with KWS's community office in Voi. During his career, he and his colleagues handled thousands of complaints from area residents concerning wildlife, many of them involving lions. Mukabane grew so familiar with the lions living on the fringes of Tsavo that he became expert in finding and eliminating problem animals. Over a fifteen-year period, Mukabane claimed to have shot and killed 223 lions and trapped and relocated hundreds of others, making him one of the most experienced animal control officers in the force. His untimely death in a motor vehicle accident was truly tragic, for it claimed one of Kenya's most knowledgeable experts on a pivotal societal problem.

Is it ever justifiable for conservation officials to kill the animals they are sworn to conserve? Although relocating problem animals seems always to be a more humane course, it is a costly one, with variable effectiveness. Problem individuals must first be captured, usually in traps, and trapped individuals must be reliably identified as problematic. Finding a place to relocate them is difficult too. Despite Tsavo's size, the number of lions in the park ought to be at equilibrium, controlled by the size of prey populations and lion territories. Any additions will entail losses somewhere else. Worse, in the case of social carnivores, if translocated individuals disrupt social groups, provoking territorial fights or infanticide, the park might lose more lions than are saved by translocation. There is also no guarantee that translocation will solve the original problem. In one study, three marauding lions were translocated and released 120 miles from their point of capture in the Kalahari Desert. Three days later, they were captured 85 miles away, and were then returned and released at the same point. A week later they had returned to the point of original capture (Eloff 1973).

Finally, the drama and stakes of animal control are heightened when dealing with man-eaters. Kruger National Park's resident scientist Ian Whyte has been quoted as saying, "It is generally accepted that lions avoid man because they have a great respect for him. This fear disappears, however, after a lion has killed a human for the first time. Lions that are held in captivity and relocated in the wild are among the most dangerous animals in the world, purely because they have lost their fear of humans" (Anonymous 1997, 11).

How Lions Fit into Conservation Plans

Conserving an entire ecosystem is inherently a complex and multifaceted task. It is plainly impossible to track all species simultaneously or to ensure that each is prospering under a given park management plan. Commonly, park managers adopt a far simpler alternative—identify one or a few species that can be used as surrogates or proxies for the entire ecosystem. The ecosystem remains the ultimate target of our conservation efforts, but focusing on individual species permits us to gauge the success of management efforts.

Different kinds of species are selected to achieve different conservation goals, as University of California wildlife biologist Tim Caro has discussed. *Umbrella species* are used to determine how big protected areas must be, because these species have the largest home ranges. If a park includes sufficient habitat for a viable population of umbrella species, then it should contain more than enough space for more modestly ranging species. *Indicator species* are used variously to identify areas of exceptional species richness or ecosystem integrity or as indices of the population sizes of other species. *Flagship species* are used to raise public awareness or attract funding to conservation; they may be useful for natural land management, for biological reintroductions, or even for captive breeding (Caro 2000). Flagship species are invariably drawn from "charismatic megafauna," animals like mountain gorillas, giant pandas, and eagles. The attention and pathos they elicit make them viewer and visitor favorites in television documentaries, zoos, and museums.

So where do lions fit in this context? Lions are large apex predators that require enormous amounts of land and sizeable prey populations to conserve them. Accordingly, they qualify well as umbrella species. Lions are also highly charismatic organisms—recent challenges to lion conservation from canine distemper and bovine tuberculosis elicited immediate attention and galvanized international action to treat them, but these actions had little impact ecosystem-wide, so lions don't really serve as a flagship species. Lions are sufficiently widespread, adaptable, and tolerant of degraded habitats that they have limited value as biological indicators, yet even here they can be useful. Poaching and habitat deterioration quickly depress populations of ungulates below the threshold needed for maintaining social carnivores. For example, healthy populations of leopards remain in the Rift Valley around Lake Naivasha, despite its relatively dense human settlements. However, lions have virtually disappeared from that region, having been effectively squeezed out of shrinking natural food webs.*

During the twentieth century, lions acquired exaggerated importance to the Kenyan communities in which they reside, for perhaps an unexpected reason: positive economic value. Wildlife conservation in Kenya is a national responsibility, overseen and executed by the Kenya Wildlife Ser-

*It should be noted, however, that lions have fared better than another social hunter. The African hunting dog is even more imperiled by human development and perhaps offers a better indicator of ecosystem health and integrity than lions.

vice, which manages both wildlife and parks and reserves. Revenues for these services are largely obtained from park entrance fees, which are modestly priced for Kenyan residents but substantial—amounting to twenty to thirty dollars per day—for foreign nationals. Basically, save for some ad hoc special-purpose grants, the fees generated by park-goers and animal watchers fuel all of the nation's park and animal management plans. And although managers may assign equivalent values to each of the animal species in the park, "some animals are more equal than others" when it comes to fund-raising, as George Orwell observed in another context. During his term as KWS Director, conservation ecologist David Western told me that exit interviews at Kenya's parks revealed that visitors spent 60 percent of their time in Kenyan parks either looking *at* lions or looking *for* them. Clearly, KWS must manage its lion populations carefully to maintain park revenues, especially when a park is as popular as Tsavo. From 1991 to 2002, an average of 130,000 visitors entered Tsavo East each year, and Tsavo West received nearly as many guests (S. M. Kasiki, personal communication).

Dangers to Park Visitors

While lions are a principal attraction for tourism, they and other wild mammals pose a continual threat to humans, even those drawn specifically to see them. Wildlife attacks on visitors are often sensational, garnering international headlines. As a result, they can have chilling effects on tourism. When a British youth was attacked and killed by lions in a Zimbabwe park in 1999, it was widely covered by the news media. The ensuing potential for depressed tourism led the Minister for Environment and Tourism to vow to keep the parks "havens of peace and safety" (Anonymous 1999). Keeping this promise may be difficult, as Matusadona National Park may support as many as 350 lions, one of the largest concentrations in Africa.

Surprisingly little is known about park visitors' risk of injury and death. One study determined the incidence of fatal and nonfatal attacks on tourists by wild mammals in South Africa, scouring records from South African newspapers over a ten-year period (January 1988 to December 1997). Throughout the decade, seven tourists were killed and fourteen visitors suffered nonfatal attacks. Most attacks, including three of the four fatal ones by lions, resulted from tourist ignorance of animal behavior or flagrant disregard of park rules—tourists had actually approached the lion prides on foot for a better view! The fourteen nonfatal attacks included five by hippos, three by buffalo, two by rhinos, and one each by a lion, a leopard, a zebra, and an elephant in musth.* Only the last occurred while the visitor was in a motor vehicle (Durrheim & Leggat 1999). These records suggest it is remarkably safe to visit African parks to see wildlife; most accidents occurred as a result of uninformed and easily avoidable mistakes on the part of the visitors.

Competing Models for Conservation

Important as they are to Kenya's conservation planning, lions are also significant components in another conservation model that is widely embraced in Southern Africa. This one depends on sport hunting and other forms of commercial utilization of wildlife. Adherents argue that by assessing usage fees for sport hunting or commerce, governments can provide the means of properly managing and conserving animals. As one of the "Big Five" African big-game species that serious sportsmen must shoot before completing their scorecards, lions are prized trophies for the sports hunter. In South Africa, where game ranching is legal, forty-five to fifty ranchers maintain about twenty-five hundred lions in cages or other enclosures on their properties, licensing the harvest of these animals through controlled hunts (Macleod 2002).

Male lions, especially those with substantial manes, are highly prized by sportsmen. Where hunting of free-living lions is permitted, as in Zambia's Luangwa Valley, selective harvest of mature male lions may disrupt their social organization (Yamazaki 1996), opening prides with cubs to takeovers by infanticidal successors. Dereck Joubert has estimated that harvesting a prime male lion may create an ecological vacuum and social unrest that eventually claims the lives of at least ten lions (Macleod 2001).

The different interests of sportsmen and conservationists recently collided in Botswana, where a government-imposed ban on trophy hunting of lions drew international criticism. A lobbying group from Safari Club International that included former president George H. W. Bush, former vice president Dan Quayle, and General Norman Schwarzkopf urged Botswana to scrap its one-year ban on lion hunting. The 2001 ban prevented the issuing of permits for a quota of fifty-three male lions, costing the nation's safari industry an estimated $5 million and depriving the government of about $100,000 in fees. Curiously, the ban was imposed not to prevent or limit the high collateral losses of lions due to trophy hunting but to mollify ranchers. Prior to the ban, safari operators had been authorized to harvest protected species on government hunting reserves at great prof-

*A physiological condition of heightened aggression and sexual readiness.

its, while farmers and ranchers were being prohibited from controlling "problem lions" decimating their stocks on private lands (Macleod 2001).

Reconciling the economics of hunting with the importance of conserving lions is essential, and scientists must offer impartial guidance. The African Lion Working Group (ALWG) of the World Conservation Union (IUCN) is currently exploring ways to ensure that trophy hunting is conservation-oriented, so that quotas do not harm the viability of lion populations. Peter Jackson, the editor of *Cat News,* urged that "hunting organisations and individual hunters must now cooperate with conservationists to ensure the lion's future" (1997, 1). Recently, the ALWG took a controversial stand in the current controversy in Botswana in support of limited hunting. Its chair, Sarel van der Merwe, stated, "We would urge that the solution to lion/livestock conflict lies not simply in a hunting ban on lions but in the implication of a number of management strategies aimed at lessening the conflict while maintaining sustainable lion populations" (African Lion Working Group 2001). Simply to ban hunting in an area where lions kill substantial numbers of livestock does not represent a satisfactory resolution to conflicts with wildlife.

The tiger's recent history identifies the challenges ahead. Trophy hunting hastened the collapse of tiger populations already fragmented by habitat conversion. Three subspecies of tigers have already gone extinct during the last century,* and only five thousand to seventy-five hundred tigers remain (Jackson 1997).† Although hunting has now been banned altogether, tigers are regularly poached to support the illegal medicine market in Asia, where skins, bones, and organs are highly coveted and utilized. Because there is no such uncontrolled commercial market for lions and their parts, there is hope for their preservation.

Prescriptions for the Future

In a real sense, the people surrounding the world's great conservation areas live on the front lines of the battle for conservation. It is they who must suf-

*The Bali tiger, *Panthera tigris balica,* went extinct in the 1940s, the Caspian tiger, *P. tigris virgata,* vanished in the 1970s, and the Javan tiger, *P. tigris sondaica,* disappeared in the 1980s (Jackson 1997).

†The populations of different geographic races of *P. tigris* for 1993 to 1997 are estimated by in-country experts as follows (minimum/maximum): Bengal tiger (*P. tigris tigris*), 3,060/4,735; Siberian tiger (*P. tigris altaica*), 437/506; South China tiger (*P. tigris amoyensis*), 20/30; Sumatran tiger (*P. tigris sumatrae*), 400/500; Indochinese tiger (*P. tigris corbetti*), 1,180/1,790 (Jackson 1997).

fer trespassing elephants and lion attacks, they who must sustain the economic damages and emotional tragedies. It is their understanding that is strained and their tolerance that is tested. The outcome of conservation planning depends ultimately on their day-to-day reactions to "living with wildlife." It is almost axiomatic that conservation will succeed only where and when humans perceive greater benefits to sustaining wildlife than to eliminating it. Such observations emphasize the importance of public education programs to wildlife organizations such as the Kenya Wildlife Service and to research institutions such as The Field Museum.

The challenge of preserving lions in Tsavo and other reserves like it can therefore be met best by focusing on the following questions: Which activities and stakeholders are adversely impacted by lions? Are there environmental or management conditions that ameliorate these effects? Do wildlife managers have effective means of disseminating information about risks, and can their assistance be secured as necessary to eliminate or redress losses? My colleagues and I have recently begun a research program to provide the KWS and the residents of the greater Tsavo ecosystem with answers to these questions. Ultimately, we hope to understand the lion-human conflict well enough to recommend effective management strategies that will maintain this top predator as an essential component of this ecosystem.

The Earthwatch Institute fosters collaboration between scientists and volunteers in the interests of enhancing nature's sustainability. Specializing in funding projects that address both human and wildlife subjects, Earthwatch was a natural funding source for our investigations. My collaborators bring conceptual breadth and experience to the project: Samuel Kasiki, a KWS scientific officer and expert in animal-human conflict and wildlife management in Tsavo, and Roland Kays, a behavioral ecologist and molecular geneticist experienced with both radiotelemetry and carnivores. We have established a field site on Taita Ranch (across the Nairobi-Mombasa Highway from Buchuma Gate of Tsavo East National Park), where we are studying lions on former ranchlands now being managed as part of a nature conservancy. Through year-round monitoring and observation of radio-collared lions, we hope to develop a detailed knowledge of their annual cycle (Plates 10 and 11). We are also documenting prey density and availability throughout the year, food habits as observed and documented in fecal samples, and lion group composition and size.

As I write this, we have just completed the first field season of observations, so only preliminary results are available. Nevertheless, some surprising results are apparent. The Earthwatch teams seem to be documenting seasonal variation in lion group size, and there is parallel variation in prey type. Animal control officer John Hunter was the first to note that Tsavo lions tend to wander during the rainy season and to escalate their attacks on livestock. During the dry season, concentrations of game around permanent water are predictable features of Tsavo's ecology. However, the dispersion of game with the onset of rains causes lion prides to fission, so that one rarely sees more than a pair together and they forage over less familiar haunts. At such times, lions may more frequently attack livestock and humans (Hunter 1952). Our new analyses of cattle depredation by lions on Taita and Rukinga ranches show that the greatest losses of livestock to lions occur during the wettest months. Lions take fewer cattle during the predictable four-month drought that stretches from June to September, but their attacks increase as the rains arrive (Patterson et al., in press). This offers important information to livestock managers in this region.

It is curious that livestock is most vulnerable to lions in Tsavo during the rains, because in Masai Mara, livestock is most secure from lion depredations during that period (Karani 1994). The explanation may be that migratory herds of zebra and wildebeest arrive in the Mara after the onset of rains to graze on the new grass, so that lions have an abundance of native game. Lions in Masai Mara face seasonal food shortages during the dry season, when migratory herds travel to the Serengeti Plains of Tanzania; it is then that they resort to raiding livestock. Therefore, in both regions of Kenya, lions take more livestock during the season of local food scarcity. However, this season differs in the context of each region's ecological cycles. Currently, a major study headed by Laurence Frank and Rosie Woodroffe of the University of California has the goal of understanding and resolving human-wildlife conflicts on Kenya's Laikipia Plateau, which offers lions and other predators yet another regional context of climatic variation and prey availability (Frank 1998).

For proper management, it is imperative that wildlife managers appreciate and understand such variation in local ecologies. Such differences spell the success or failure of any plan developed from them. If they can limit losses to predators and enhance the profitability of their operations, cattle ranchers in Tsavo need not consider unsustainable alternatives such as illegally poisoning lions or converting their ranchlands to cultivated sisal fields. The arid rangelands are an environment in which lions, humans, and a host of other savanna-dependent species should be able to coexist. Science must furnish the answers that will foster this understanding and permit this coexistence.

It is remarkable that Tsavo, once considered a virtual wasteland, should now constitute one of safest and the most important wildlife preserves in

East Africa. It is ironic that the region that deprived the lion of its majestic mane should offer him something much more valuable instead—prospects for a future. And it is both sobering and inspiring that we should inherit the stewardship of both lions and their landscape, for their fates will surely be decided within our lifetimes and by our hands.

Acknowledgments

The genesis of this book lay in countless conversations with Field Museum colleagues Chap Kusimba, Tom Gnoske, and Julian Kerbis Peterhans in 1997 and 1998. After several unsuccessful attempts, and with the superb assistance of the Kenya Wildlife Service, Gnoske and Kerbis Peterhans had succeeded in re-locating a small cave that Col. John Henry Patterson had dubbed the "Man-eaters' Den," where he supposed the lions had feasted on human remains. The find had aroused the interest of the Kenya Wildlife Service's former director David Western and his assistant director, Bella Ochola Wilson, who sought the partnership of the Field Museum in developing a visitors' center in Tsavo. The principal goals of the center would be to educate visitors about the Tsavo parks and their crucial role in East African conservation efforts.

A visitors' center in Tsavo was an interesting proposition—it would be located in one of Kenya's oldest, largest, and most visited national parks and accessible via a major highway linking Kenya's two largest cities. And it would help to answer innumerable questions about the area's notorious history, especially those revolving around its lions. The box-office success of the Paramount Pictures film *The Ghost and the Darkness* had just confirmed the world's continuing interest in this epic story. The drama's potential to engage visitors long enough and deeply enough to educate them about the story's environmental context was obvious to all of us. The cave, which united the disparate research interests of Kusimba, Gnoske, and Kerbis Peterhans, also promised to be a lodestone for their research in taphonomy, archeology, and natural history. They invited and encouraged me to initiate complementary studies of Tsavo's lions.

Over the ensuing year, Kusimba, Gnoske, Kerbis Peterhans, and I met repeatedly and at length to discuss research and exhibition possibilities at Tsavo. We collaborated in writing a synopsis of our research interests, which appeared in *Natural History* magazine (Kerbis Peterhans 1998); jointly wrote a proposal for fieldwork and an accompanying television documentary (Kusimba 2001); and worked with Dan Brinkmeier of the museum's Education Department on a larger, integrated plan for research, education, and exhibition that never bore fruit (Patterson 1998). The group even reviewed the prospectus for this book, in its various avatars. Administrative changes at KWS and shifting interests in our group caused us to abandon our collaboration in 2000. Since that time, Chap Kusimba has worked with his wife, Sibel Barut Kusimba, on an ambitious synopsis of the region's prehistory, and he has graciously permitted me access to some of his preprints and all of his expertise. Meanwhile, Tom Gnoske and Julian Kerbis Peterhans have pursued their own studies of lions, man-eating, and taphonomy. Here I acknowledge all three for encouraging my interest in the region's fascinating history and for discussions on the relationships between lions and other topics at Tsavo. I thank Tom in particular for calling my attention to some important literature records I might otherwise have missed. While they deserve credit for helping to frame my interests in those early discussions, any omissions or errors in scholarship are exclusively mine.*

My own investigations of Tsavo's lions owe much to my collaborators. Roland Kays deserves first mention. He helped initiate the field studies as a postdoctoral fellow at the Field Museum and has provided energy, insight, and enthusiasm at every turn. Roland's broad interests, determination, and collegial spirit have made our collaboration productive and rewarding. Although Samuel Kasiki, Kenya Wildlife Service's Resident Scientist at Tsavo East, joined our group only in 2001, he has quickly become indispensable and led "the lion's share" of the Earthwatch volunteer teams in 2002. His quiet efficiency and thorough professionalism make our international collaboration work. Jean Dubach, a conservation geneticist at Brookfield Zoo graciously allowed us to join in her ongoing analyses of African lion genetics long after they had begun. The scope of her studies with Mike Briggs on southern African lions has greatly enlarged the footprint of our genetic studies and helped us to place the Tsavo lions in a continent-wide

*Unfortunately, this chronology caused two other works on the lions of Tsavo to fall outside the purview of this book, for the simple reason that all were prepared independently over the same time period. Philip Caputo and I announced our respective book projects in January 2000, while Kerbis Peterhans and Gnoske communicated their plans for a technical article on the subject later that year.

picture. I also owe much to Ellis (Skip) Neiburger, whose expertise with teeth carried my dental investigations into pathology and behavior; I thank Bill Turnbull for catalyzing this partnership with his infectious curiosity and collegiality. Tom Gnoske was a member of the lion team in its early days and shared ideas and speculations on the significance of maneless lions. Sarah Lansing, Barbara Harney, and Dan Patterson assisted Roland and me with fieldwork, and Sarah additionally cleaned and measured "problem" lions in the Tsavo Research Centre.

Maria Collavicenzo and Harvey Golden helped compile and organize relevant literature. Harvey's thoughtful scrutiny and insightful analyses of these articles fueled many meaningful discussions that are reflected in various sections of these pages. And despite the excellence of the Museum's General Library, I continually sought and received assistance of all sorts from Stephanie Stephens, our Inter-Library Loan officer. Clara Simpson drafted the precipitation contour map, and her skill is much appreciated. I also need to thank Chap Kusimba, curator of African Anthropology and Ethnography at the Field Museum and director of the museum's cultural investigations at Tsavo. His enduring vision and tangible support for an integrated program of natural and cultural history set an inspiring example of both collegiality and friendship.

My colleagues and I have benefited from the help and generosity of numerous organizations and individuals. A principal debt is owed the Kenyan Wildlife Service for permission to work in their world-renowned parks and for the expert help and enthusiastic hospitality of its officers and rangers. Special thanks go out to Dr. Richard Bagine, Dr. Elizabeth Wambwa, Dr. Francis Gakuya, John Muhanga, Natali Kio, Ali Salim, Walter Njuguma, and Patrick Mulandi at Headquarters; Robert Mwasya, John Kimani, Joseph Kisio, and Charles Upiche at Manyani; Peter Leitoro, William Mukabane, James Kagui, Alex Mwazo and Barbara McKnight at Tsavo East; and James Isiche, Mr. Kaisha, and Dr Samuel Andanje at Tsavo West. At the National Museums of Kenya, I wish to thank Dr. Karege-Munene, Nina Mudida, Mwabi Ogeto, Kiberenge Musombi, Yakub Dahiye, and especially Chui Ng'ang'a for their assistance; at the Natural History Museum, London, Paula Jenkins and Daphne Hill; and at the National Museum of Natural History (Smithsonian Institution), Helen Kafka and Linda Gordon.

Funding for the initial stages of this work was provided by the Barbara E. Brown Fund for Mammal Research, the Street Mammal Expeditionary Fund, and the Marshall Field III Fund (all at the Field Museum) and by the Eli Lilly Foundation, the Explorers Club, the Negaunee Foundation, and Jake and Catherine Jacobus. The Negaunee Foundation covered our vehicle expenses at the very onset of our Tsavo research, when such support was critical. The Brown Fund facilitated my initial collaboration with Roland Kays, underwriting part of his postdoctoral fellowship, and the Jacobus family provided the support for the fieldwork that has formed the foundation of our work in Tsavo. I am extremely grateful to all three sources, and it is a special pleasure to salute the generosity and imagination of the people involved: Robin and Richard, Barbara and Roger, and Catherine and Jake.

Currently, our field and laboratory studies are supported by the Committee on Research and Exploration, National Geographic Society (Committee on Research and Exploration Grant #7208-02) and by the Earthwatch Institute (Contract #5123). We are grateful for the collaboration of East African Ornithological Safaris and Steve Turner in hosting the Earthwatch teams and coordinating their activities, and to Pamela and Hervé De Maigret for help in buying the satellite collar we placed on Romeo. The expertise of Galla Camp's staff, especially Paul Kabochi but also Sammy Kabaiko, Simon Wamuya Wanjohi, Andrew Karanja, Allan Kiigi, Peter Gichimu, Octavian Mkuria, and Joseph Nambala, is gratefully acknowledged. Paul's tragic death on 8 March 2003, at the foot of the *kopje* on which our camp is situated, has left us all bereft of his curiosity, wit, and integrity—for the people who loved him, Paul will forever ride on Galla's fair breezes.

Special thanks are due Barbara Harney and Robert D. Martin, who read the entire manuscript. At McGraw-Hill, I am grateful to Griffin Hansbury for his initial interest in the project, as well as to Anthony Sarchiapone and Philip Ruppel for their help in bringing the book to a satisfactory conclusion. Ruth Mills, Amy Fass and Peter McCurdy offered a host of editorial suggestions along the way that helped focus, shorten, and clarify the work.

References

CHAPTER 1: The Reign of Terror: The Lions of Tsavo Attack

Astley Maberly, C. T. 1963. *The Game Animals of Southern Africa*. Thomas Nelson and Sons, Johannesburg.

Beard, P. H. 1988. *The End of the Game: The Last Word from Paradise*. Thames & Hudson Ltd, London.

Carlozo, L. 1996. Field maneuvers. *Chicago Tribune* for 14 October 1996, Tempo Section, pages 1,5.

Davies, D. C. 1926. Annual report of the Director to the Board of Trustees for the year 1925. *Field Museum of Natural History. Report Series*, 6: iv + 390–514.

Foran, R. *The Kima Killer*. East African Railways, Nairobi, 1961.

Guggisberg, C. A. W. 1961. *Simba, the Life of the Lion*. Howard Timmins, Capetown, South Africa.

Hardy, R. 1965. *The Iron Snake*. G. P. Putnam's Sons, New York.

Hill, M. F. 1949. *Permanent Way: The Story of the Kenya and Uganda Railway*. Vol. 1. East African Literature Bureau, Nairobi.

Kays, R. W., and B. D. Patterson. 2002. Mane variation in African lions and its social correlates. *Canadian Journal of Zoology*, 80: 471–478.

Miller, C. 1971. *The Lunatic Express: An Entertainment in Imperialism*. Westlands Sundries, Nairobi.

Patterson, J. H. 1907. *The Man-eaters of Tsavo and Other East African Adventures*. Macmillan, London.

Patterson, J. H. 1925. *The Man-eating Lions of Tsavo*. Zoology: Leaflet 7, Field Museum of Natural History, Chicago.

Preston, R. O. n.d. *The Genesis of Kenya Colony: Reminiscences of an Early Uganda Railway Construction Pioneer*. The Colonial Printing Works, Nairobi.

Roosevelt, T. 1909. *African Game Trails. An Account of the African Wanderings of an American Hunter-Naturalist*. Charles Scribner's Sons, New York.

Roosevelt, T., and E. Heller, 1914. *Life-Histories of African Game Animals.* Vol. 1, Charles Scribner's Sons, New York.

Thomson, J. 1885. *Through Masai Land: A Journey of Exploration among the Snowclad Volcanic Mountains and Strange Tribes of Eastern Equatorial Africa.* Sampson Low, Marston, Searle, & Rivington, London.

CHAPTER 2: The Terror Continues: Man-eating Lions Today

Hosek, W. in prep. *The Man-eater of Mfuwe.*

Prothero, W. 1996. The reluctant hero and the man-eater. *Safari*, 29: 122–124.

Treves, A., and L. Naughton-Treves. 1999. Risk and opportunity for humans coexisting with large carnivores. *Journal of Human Evolution*, 36: 275–282.

Vosper, R. 1998. Museum receives largest man-eating lion on record. *In the Field*, 69: 6.

Yamazaki, K. 1996. Social variation of lions in a male-depopulated area in Zambia. *Journal of Wildlife Management*, 60: 490–497.

Yamazaki, K., and T. Bwalya. 1999. Fatal lion attacks on local people in the Luangwa Valley, Eastern Zambia. *South African Journal of Wildlife Research*, 29: 19.

CHAPTER 3: Killing Behavior and Man-eating Habits

Akeley, C. E. 1920. *In Brightest Africa.* Garden City Publishing Company, Garden City, NY.

Anderson, K. 1955. *Nine Man-eaters and One Rogue.* E. P. Dutton & Co., New York.

Anonymous. 1989. Man-eating lions killed. *Cat News*, 12.

Anonymous. 1999. Zimbabwe apologises over Briton mauled by lions. in *Daily Mail* & *Guardian* for 4 August 1999, Johannesburg, South Africa.

Astley Maberly, C. T. 1963. *The Game Animals of Southern Africa.* Thomas Nelson and Sons, Johannesburg.

Elliot, D. G. 1897. Lists of mammals from Somali-land obtained by the museum's East African expedition. *Field Columbian Museum: Zoological Series*, 1: 109-155.

Eloff, F. C. 1973. Ecology and behavior of the Kalahari lion, 90-126 in *The World's Cats* (ed. R. L. Eaton), World Wildlife Safari, Winston, Oregon.

Fitzsimons, F. W. 1919. *The Natural History of South Africa*, Vol. 1. Longmans, Green and Co, London.

Guggisberg, C. A. W. 1975. *Wild Cats of the World.* Taplinger Publishing Company, New York.

Hunter, J. A. 1952. *Hunter.* Harper & Brothers, Publishers, New York.

Johnson, M. 1928. *Safari: A Saga of the African Blue.* G. P. Putnam's Sons, New York.

Kitchener, A. 1991. *The Natural History of the Wild Cats.* Comstock Publishing Associates, Ithaca, New York.

Leyhausen, P. 1979. *Cat Behaviour: The Predatory and Social Behavior of Domestic and Wild Cats.* Garland STPM Press, New York.

Patterson, J. H. 1907. *The Man-eaters of Tsavo and Other East African adventures.* Macmillan, London.

Roosevelt, T., and E. Heller. 1914. *Life-histories of African Game Animals.* Charles Scribner's Sons, New York.

Rushby, G. G. 1965. *No More the Tusker.* W. H. Allen, London.

Schaller, G. B. 1972. *The Serengeti Lion; A Study of Predator-Prey Relations.* University of Chicago Press, Chicago.

Selous, F. C. 1881. *A Hunter's Wanderings in Africa.* Richard Bentley & Son, London.

Swayne, H. C. G. 1895. *Seventeen Trips through Somaliland. A Record of Exploration & Big Game Shooting, 1885 to 1893.* Rowland Ward and Co., London.

Taylor, J. 1959. *Maneaters and Marauders.* A.S. Barnes and Co., New York.

Van Orsdol, K. G. 1984. Foraging behaviour and hunting success of lions in Queen Elizabeth National Park, Uganda. *African Journal of Zoology*, 22: 79-99.

Waterfield, G. 1966. First footsteps in East Africa by Sir Richard Burton. in *First Footsteps in East Africa* by Sir Richard Burton, London.

CHAPTER 4: Why Do Lions Kill People?

Anderson, K. 1955. *Nine Man-eaters and One Rogue.* E. P. Dutton & Co., New York.

Arenstein, J. 1998. The outcast leopard that turned killer. *Daily Mail & Guardian* for 27 August 1998, Johannesburg, South Africa.

Astley Maberly, C. T. 1963. *The Game Animals of Southern Africa.* Thomas Nelson and Sons, Johannesburg.

Balestra, F. A. 1962. The man-eating hyaenas of Mulanje. *African Wildlife*, 16: 25–27.

Blau, H. 1961. The Black Art of Robert Lavigne. in *The Black Art of Robert Lavigne*, San Francisco.

Bodry-Sanders, P. 1991. *Carl Akeley: Africa's Collector, Africa's Savior.* Paragon House, New York.

Chellam, R. 1996. Lions of the Gir Forest. *Wildlife Conservation*, 99: 40.

Corbett, J. E. 1946. *Man-eaters of Kumaon.* Oxford University Press, London.

Corbett, J. 1955. *The Temple Tiger, and More Man-eaters of Kumaon.* Oxford University Press, New York.

Cowie, M. 1966. *The African Lion.* Golden Press, New York.

Denis, A. 1964. *Cats of Africa: The African Lion.* Houghton Mifflin Company, Boston.

Dobson, A. 1995. The ecology and epidemiology of rinderpest virus in Serengeti and Ngorongoro Conservation area. In *Serengeti II: Dynamics, Management and Conservation of an Ecosystem*, edited by A. R. E. Sinclair and P. Arcese, 485–505. University of Chicago Press, Chicago.

Edey, M. 1968. *The Cats of Africa.* Time-Life Books, New York.

Eloff, F. C. 1973. Ecology and behavior of the Kalahari lion. In *The World's Cats,* edited by R. L. Eaton, 90–126. World Wildlife Safari, Winston, OR

Fitzsimons, F. W. 1919. *The Natural History of South Africa*, Vol. 1. Mammals Longmans, Green and Co., London.

Graham, A., and P. H. Beard. 1973. *Eyelids of Morning: The Mingled Destinies of Crocodiles and Men.* Chronicle Books, San Francisco.

Hardinge, A. Report on the British East Africa Protectorate for the year 1897–98. Her Majesty's Stationery Office, London, 1899.

Guggisberg, C. A. W. 1975. *Wild Cats of the World.* Taplinger Publishing Company, New York.

Hardy, R. 1965. *The Iron Snake.* G. P. Putnam's Sons, New York.

Homer. *The Iliad*, translated by E. V. Rieu, 1966. Penguin Books, Baltimore, MD.

Hunter, J. A. 1952. *Hunter.* Harper & Brothers, Publishers, New York.

Kasperson, J. X., R. E. Kasperson, II, B. L. T., eds., 1995. Regions at risk: comparisons of threatened environments. United Nations University Press, Tokyo.

Kerbis Peterhans, J. C., C. M. Kusimba, T. P. Gnoske, S. A. Andanje, and B. D. Patterson. 1998. Man-eaters of Tsavo. *Natural History*, 107: 12–14.

Kingdon, J. 1974. *East African Mammals. An Atlas of Evolution in Africa.* Academic Press, London.

Kruuk, H., 1972. *The Spotted Hyena*. University of Chicago Press, Chicago.

Kusimba, C. M. In prep. Early European accounts of the peoples and cultures of Tsavo. In *Early European accounts of the peoples and cultures of Tsavo*, Chicago.

Matthiessen, P., 1972. *The Tree Where Man Was Born.* E. P. Dutton & Co., New York.

Melland, F. H. 1923. *In Witch-bound Africa, an Account of the Primitive Kaonde Tribe & Their Beliefs*. Seeley Service & Co. Limited, London.

Nassau, R. H. 1904. *Fetichism in West Africa: Forty Years' Observation of Native Customs and Superstitions*. Charles Scribner's Sons, New York.

Neiburger, E. J., and B. D. Patterson. 2000a. Man eating lions…a dental link. *Journal of the American Association of Forensic Dentists* 24:1-3.

Neiburger, E. J., and B. D. Patterson. 2000b. The man-eaters with bad teeth. *New York State Dental Journal* 66:26-29 + cover.

Neiburger, E. J., and B. D. Patterson. 2002. A forensic dental determination of serial killings by three African lions. *General Dentistry* 50:40-42.

Nowak, R. M. 1999. *Walker's Mammals of the World.* 6th ed. Johns Hopkins University Press, Baltimore, MD.

Oka, R., C. M. Kusimba, and P. F. Thorbahn. In prep. *East African Ivory and Indian Socio-politics: The Role of the Indian Subcontinent in the Development of the Ivory Trade of East Africa in the Indian Ocean Economic Complex.*

Patterson, B. D., and E. J. Neiburger. 2001. Lion with a sore tooth. *Nature Australia* 26:12.

Patterson, B. D., E. J. Neiburger, and S. M. Kasiki. 2003. Tooth breakage and dental disease as causes of carnivore-human conflicts. *Journal of Mammalogy* 84:190-196.

Patterson, J. H., 1986. *The Man-eaters of Tsavo*. 1st edn. St. Martin's Press, New York.

Polak, B. 1986. *Rinderpest and Kenya in the 1890s*. M.A. thesis, Northwestern University, Evanston, IL

Rabinowitz, A. 1986. *Jaguar: Struggle and Triumph in the Jungles of Belize*. Arbor House, New York.

Ramdhani, N. 1989. The effects of climate and disease on African farming in Natal 1895–1905. *The South African Journal of Economic History*, 4: 79–90.

Roberts, A. F. 1986. "Like a roaring lion": Tabwa terrorism in the late nineteenth century. In *Banditry, Rebellion and Social Protest in Africa*, edited by D. Crummey, pp. 65-86. Heinemann, Portsmouth, NH.

Roosevelt, T. 1909. *African Game Trails. An Account of the African Wanderings of an American Hunter-Naturalist*. Charles Scribner's Sons, New York.

Roosevelt, T., and E. Heller. 1914. *Life-Histories of African Game Animals*. Charles Scribner's Sons, New York.

Rushby, G. G. 1965. *No More the Tusker*. W. H. Allen, London.

Schneider, H. K. 1982. Male-female conflict and lion men of Singida. In *African Religious Groups & Beliefs*, edited by O. Ottenberg, 95–109 Folklore Institute, Delhi, India.

Spinage, C. A. 1973. A review of ivory exploitation and elephant population trends in Africa. *East African Wildlife Journal*, 11: 281–289.

Stevenson-Hamilton, J. 1947. *Wildlife in South Africa*. Cassel & Co., London.

Taylor, J. 1959. *Maneaters and Marauders*. A. S. Barnes and Co., New York.

Van Valkenburgh, B. 1988. Incidence of tooth breakage among large, predatory mammals. *The American Naturalist*, 131: 291–300.

Van Valkenburgh, B., and F. Hertel. 1993. Tough times at La Brea: tooth breakage in large carnivores of the late Pleistocene. *Science*, 261: 456–459.

CHAPTER 5: Lion Biology: Evolution and Geographic Distribution

Anonymous. 1998. Bovine tuberculosis in South Africa's lions. *Cat News*, 29: 21–22.

Ansell, W. F. H. 1960. *Mammals of Northern Rhodesia*. The Government Printer, Lusaka, Zambia.

Anyonge, W. 1993. Body mass in large extant and extinct carnivores. *Journal of Zoology*, London, 231: 339–350.

Astley Maberly, C. T. 1963. *The Game Animals of Southern Africa*. Thomas Nelson and Sons, Johannesburg.

Ballesio, R. 1975. Étude de *Panthera (Leo) spelaea* (Goldfuss) nov. subsp. (Mammalia, Carnivora, Felidae) du gisement Pleistocene moyen des abimes de la fagne á Noailles (Correze). *Nouvelles Archives du Museum d'Histoire Naturelle de Lyon*, 13: 47–55.

Bertram, B. C. R. 1978. *Pride of Lions.* Scribner, New York.

Caputo, P. 2000. Among the man-eaters. *National Geographic Adventure,* 2(3): 74–94, 146–149.

Caputo, P. 2002. *Ghosts of Tsavo: Stalking the Mystery Lions of East Africa.* Adventure Press National Geographic, Washington, D.C.

Chellam, R. 1996. Lions of the Gir Forest. *Wildlife Conservation,* 99: 40.

Chellam, R., and A. J. T. Johnsingh. 1993. Management of Asiatic lions in the Gir Forest, India. In *Mammals as Predators,* edited by N. Dunstone and M. L. Gorman, 409–424. Clarendon Press, Oxford.

Coheleach, G. 1982. *The Big Cats: The Paintings of Guy Coheleach.* Harry N. Abrams, Inc., New York.

Dobzhansky, T. G. 1970. *Genetics of the Evolutionary Process.* Columbia University Press, New York.

Ellerman, J. R., and T. C. S. Morrison-Scott. 1951. *Checklist of Palaearctic and Indian Mammals, 1758–1946.* Printed by order of the Trustees of the British Museum, London.

Ellerman, J. R., and T. C. S. Morrison-Scott. 1966. *Checklist of Palaearctic and Indian Mammals, 1758–1946.* British Museum (Natural History), London.

Excoffier, L., and S. Schneider. 1999. Why hunter-gatherer populations do not show signs of Pleistocene demographic expansions. *Proceedings of the National Academy of Sciences of the USA,* 96: 10597–10602.

Fitzsimons, F. W. 1919. *The Natural History of South Africa.* Vol. 1. Longmans, Green and Co, London.

Flynn, J. J., and M. A. Nedbal. 1998. Phylogeny of the Carnivora (Mammalia): congruence vs. incompatibility among multiple data sets. *Molecular Phylogenetics and Evolution,* 9: 414–426.

Gnoske, T., and J. Kerbis Peterhans. 2000. Cave lions: the truth behind biblical myths. *In the Field,* 71: 2–6.

Guggisberg, C. A. W. 1961. *Simba, the Life of the Lion.* Howard Timmins, Capetown, South Africa.

Guggisberg, C. A. W. 1975. *Wild Cats of the World.* Taplinger Publishing Company, New York.

Hall, E. R. 1981. *The Mammals of North America.* 2d ed. Vol. 2. John Wiley and Sons, New York.

Hallgrímsson, B., and V. Maiorana. 2000. Variability and size in mammals and birds. *Biological Journal of the Linnean Society,* 70: 571–595.

Harper, F. 1945. *Extinct and Vanishing Animals of the Old World.* Special Publication Vol. 12, American Committee for International Wildlife Protection, New York Zoological Park, New York.

Hast, M. H. 1986. The larynx of roaring and non-roaring cats. *Journal of Anatomy,* 149: 221–222.

Heaney, G. F. 1943. Occurrence of the lion in Persia. *Journal of the Bombay Natural History Society,* 44: 467.

Hennig, W. 1966. *Phylogenetic Systematics.* University of Illinois Press, Urbana.

Hunter, J. A. 1952. *Hunter.* Harper & Brothers, Publishers, New York.

Jani, R. G., and P. K. Malik. 1997. Conservation of Asiatic Lion: A biotechnological approach. *Indian Journal of Forestry*, 20: 403–405.

Janis, C. M. 1993. Tertiary mammal evolution in the context of changing climates, vegetation, and tectonic events. *Annual Review of Ecology and Systematics*, 24: 467–500.

Johnson, M. 1929. *Lion: African Adventure with the King of Beasts.* G. P. Putnam, New York.

Johnson, W. E., P. A. Dratch, J. S. Martenson, and S. J. O'Brien. 1996. Resolution of recent radiations within three evolutionary lineages of Felidae using mitochondrial restriction fragment length polymorphism variation. *Journal of Mammalian Evolution*, 3: 97–120.

Johnson, W. E., and S. J. O'Brien. 1997. Phylogenetic reconstruction of the Felidae using 16S rRNA and NADH-5 mitochondrial genes. *Journal of Molecular Evolution*, 44: S98–S116.

Kurtén, B. and A. Anderson. 1980. *Pleistocene Mammals of North America.* Columbia University Press, New York.

Mattern, M. Y., and D. A. McLennan. 2000. Phylogeny and speciation of felids. *Cladistics*, 16: 232–253.

Mazak, V. 1970. The Barbary lion, *Panthera leo leo* (Linnaeus, 1758); some systematic notes, and an interim list of the specimens preserved in European museums. *Zeitschrift für Säugetierkunde*, 35: 34–44.

Mazák, V. 1975. Notes on the black-maned lion of the Cape, *Panthera leo melanochaita* (Ch. H. Smith, 1842) and a revised list of the preserved specimens. North-Holland Publishing Company, Amsterdam.

Meester, J. A. J., and H. W. Setzer. 1971 (as revised 1977). *The Mammals of Africa: An Identification Manual.* Smithsonian Institution, Washington, D.C.

Miththapala, S., J. Seidensticker, and S. J. O'Brien. 1996. Phylogeographic subspecies recognition in leopards (*Panthera pardus*): molecular genetic variation. *Conservation Biology* 10: 1115-1132.

Packer, C., and P. West. 2000. Frequently asked questions. http://www.lionresearch.org/FAQ/FAQS.html (for 15 February 2000).

Paetau, D., L. P. Waits, P. I. Clarkson, L. Craighead, E. Vyse, R. Ward, and C. Strobeck. 1998. Variation in genetic diversity across the range of North American brown bears. *Conservation Biology*, 12: 418–429.

Patterson, J. H. 1907. The Man-eaters of Tsavo and Other East African Adventures. Macmillan, London.

Peters, G., and M. H. Hast. 1994. Hyoid structure, laryngeal anatomy, and vocalization in felids (Mammalia: Carnivora: Felidae). *Zeitschrift für Säugetierkunde*, 59: 87–104.

Pocock, R. I. 1917. The classification of existing Felidae. *Annals & Magazine of Natural History*, (8)20: 329–350.

Pocock, R. I. 1931. The lions of Asia. *Journal of the Bombay Natural History Society*, 34: 638–665.

Pocock, R. I. 1939. *The Fauna of British India, Mammalia.* Vol. 1, Primates and Carnivora. Taylor & Francis, London.

Potts, R., and A. L. Deino. 1995. Mid-Pleistocene change in large mammal faunas of East Africa. *Quaternary Research*, 43: 106–113.

Rodgers, W. A. 1974. Weights, measurements and parasitic infestation of six lions from southern Tanzania. *East African Wildlife Journal*, 12: 157–158.

Roosevelt, T., and E. Heller. 1914. *Life-Histories of African Game Animals.* Charles Scribner's Sons, New York.

Russell, W. M. S. 1994. Greek and Roman monsters. Part 2. *Social Biology and Human Affairs*, 59: 1–8.

Selous, F. C. 1881. *A Hunter's Wanderings in Africa.* Richard Bentley & Son, London.

Singh, H. S. 1995. Population dynamics, group structure and natural dispersal of Asiatic lions. *The Indian Forester*, 121: 871.

Smithers, R. H. N. 1971. *The Mammals of Botswana.* The Trustees of the National Museums of Rhodesia, Salisbury.

Smithers, R. H. N. 1983. *The Mammals of the Southern African Subregion.* University of Pretoria, Pretoria.

Stevenson-Hamilton, J. 1912. *Animal Life in Africa.* E. P. Dutton, New York.

Strait, D. S., and B. A. Wood. 1999. Early hominid biogeography. *Proceedings of the National Academy of Sciences of the USA*, 96: 9196–9200.

Todd, N. B. 1966. Metrical and non-metrical variation in the skulls of Gir lions. *Journal of the Bombay Natural History Society*, 62: 507–520.

Turner, A. 1985. Extinction, speciation and dispersal in African larger carnivores, from the late Miocene to Recent. *South African Journal of Science*, 81: 256–257.

Turner, A., and M. Antón. 1997. *The Big Cats and Their Fossil Relatives: An Illustrated Guide to Their Evolution and Natural History.* Columbia University Press, New York.

Van Valkenburgh, B. 1999. Major patterns in the history of carnivorous mammals. *Annual Review of Earth and Planetary Sciences*, 27: 463–493.

von Buol, P. 2000. "Buffalo lions": A feline missing link? *Swara*, 23: 20–25.

Werdelin, L., and L. Olsson. 1997. How the leopard got its spots: A phylogenetic view of the evolution of felid coat patterns. *Biological Journal of the Linnean Society*, 62: 383–400.

Yalden, D. W., M. J. Largen, and D. Kock. 1980. Catalogue of the mammals of Ethiopia. Part 4, Carnivora. *Monitore Zoologico Italiano*, n. s. (supplement) 13: 169–272.

CHAPTER 6: Hunting and Social Behavior

Akeley, C. E. 1920. *In Brightest Africa.* Garden City Publishing Company, Garden City, NY.

Andersen, K. F., and T. Vulpius. 1999. Urinary volatile constituents of the lion, *Panthera leo. Chemical Senses*, 24: 179–189.

Ansell, W. F. H. 1960. *Mammals of Northern Rhodesia*. The Government Printer, Lusaka.

Astley Maberly, C. T. 1963. *The Game Animals of Southern Africa*. Thomas Nelson and Sons, Johannesburg.

Atmar, W. 1991. On the role of males. *Animal Behaviour*, 41: 195–205.

Ayeni, J. S. O. 1975. Utilization of water holes in Tsavo National Park (East). *East African Wildlife Journal*, 13: 305–323.

Bertram, B. C. R. 1973. Lion population regulation. *East African Wildlife Journal*, 11: 215–225.

Bygott, J. D., B. C. R. Bertram, and J. P. Hanby. 1979. Male lions in large coalitions gain reproductive advantages. *Nature*, 282: 839–841.

Elliott, J. P., and I. M. Cowan. 1978. Territoriality, density, and prey of the lion in Ngorogoro Crater, Tanzania. *Canadian Journal of Zoology*, 56: 1726–1734.

Eloff, F. C. 1964. On the predatory habits of lions and hyaenas. *Koedoe*, 7: 105–112.

Eloff, F. C. 1973. Ecology and behavior of the Kalahari lion. In *The World's Cats*, edited by R. L. Eaton, 90-126. World Wildlife Safari, Winston, Oregon.

Eloff, F. C. 1973. Water use by the Kalahari lion, *Panthera leo vernayi*. *Koedoe*, 16: 149–154.

Fitzsimons, F. W. 1919. *The Natural History of South Africa*, Vol. 1. Longmans, Green and Co., London.

Funston, P. J., M. G. L. Mills, and H. C. Biggs. 2001. Factors affecting the hunting success of male and female lions in the Kruger National Park. *Journal of Zoology, London*, 253: 419–431.

Funston, P. J., M. G. L. Mills, H. C. Biggs, and P. R. K. Richardson. 1998. Hunting by male lions: ecological influences and socioecological implications. *Animal Behaviour*, 56: 1333–1345.

Grinnell, J., and K. McComb. 1996. Maternal grouping as a defense against infanticide by males: Evidence from field playback experiments on African lions. *Behavioral Ecology*, 7: 55.

Grinnell, J., C. Packer, and A. E. Pusey. 1995. Cooperation in male lions: Kinship, reciprocity or mutualism? *Animal Behaviour*, 49: 95.

Guggisberg, C. A. W. 1961. *Simba, The Life of the Lion*. Howard Timmins, Cape Town, South Africa.

Guggisberg, C. A. W. 1975. *Wild Cats of the World*. Taplinger Publishing Company, New York.

Guthrie, R. D. 1990. *Frozen Fauna of the Mammoth Steppe: The Story of Blue Babe.* University of Chicago Press, Chicago.

Heinsohn, R., C. Packer, and A. E. Pusey. 1996. Development of cooperative territoriality in juvenile lions. *Proceedings of the Royal Society of London, Series B*, 263: 475.

Hunter, J. A. 1952. *Hunter.* Harper & Brothers, Publishers, New York.

Hunter, L. T. B. 1999. Dangerous liaisons. *Africa: Environment and Wildlife*, 7: 20–23.

Kat, P. 2000. Prides and prejudice: Challenging the accepted truths about lions' family lives. *BBC Wildlife*, 18: 28.

Kat, P. W., and C. Harvey. 2000. *Prides: The Lions of Moremi.* Smithsonian Institution Press, Washington, D.C.

Kruuk, H. 1972. *The Spotted Hyena*. University of Chicago Press, Chicago.

Legge, S. 1996. Cooperative lions escape the Prisoner's Dilemma. *Trends in Ecology and Evolution*, 11: 2.

McComb, K., C. Packer, and A. E. Pusey. 1994. Roaring and numerical assessment in contests between groups of female lions, *Panthera leo*. *Animal Behaviour*, 47: 379–387.

Natoli, E. 1990. Mating strategies in cats: A comparison of the role and importance of infanticide in domestic cats, *Felis catus* L., and lions, *Panthera leo* L. *Animal Behaviour*, 40: 183.

Packer, C., A. E. Pusey, and L. E. Eberly. 2001. Egalitarianism in female African lions. *Science*, 293: 690–693.

Packer, C., and L. Ruttan. 1988. The evolution of cooperative hunting. *The American Naturalist*, 132: 159–198.

Packer, C., D. Scheel, and A. E. Pusey. 1990. Why lions form groups: Food is not enough. *The American Naturalist*, 136: 1–19.

Pienaar, A. A. 1969. Predator-prey relationships amongst the larger mammals of the Kruger National Park. *Koedoe*, 12: 108–187.

Prins, H. H. T., and G. R. Iason. 1989. Dangerous lions and nonchalant buffalo. *Behaviour*, 108: 262–296.

Pusey, A. E., and C. Packer. 1983. Once and future kings. *Natural History*: 54–62.

Pusey, A. E., and C. Packer. 1987. The evolution of sex-biased dispersal in lions. *Behaviour*, 101: 275–310.

Rodgers, W. A. 1974. The lion (*Panthera leo*, Linn.) population of the eastern Selous Game Reserve. *East African Wildlife Journal*, 12: 313–317.

Roosevelt, T., and E. Heller. 1914. *Life-Histories of African Game Animals.* Charles Scribner's Sons, New York.

Rudnai, J. A. 1973. Reproductive biology of lions (*Panthera leo massaica* Neumann) in Nairobi National Park. *East African Wildlife Journal*, 11: 241–253.

Rudnai, J. A. 1974. The pattern of lion predation in Nairobi Park. *East African Wildlife Journal*, 12: 213–225.

Rushby, G. G. 1965. *No More the Tusker.* W. H. Allen, London.

Schaller, G. B. 1972. *The Serengeti Lion: A Study of Predator-Prey Relations.* University of Chicago Press, Chicago.

Scheel, D. 1993. Profitability, encounter rates, and prey choice of African lions. *Behavioral Ecology*, 4: 90.

Scheel, D. L. 1992. *Foraging behavior and predator avoidance: Lions and their prey in the Serengeti* (Panthera leo, Phacochoerus aethiopicus, Connochaetes taurinus, Equus burchelli, Syncerus caffer, Tanzania). Ph.D. diss., University of Minnesota, Minneapolis.

Singh, H. S. 1995. Population dynamics, group structure and natural dispersal of Asiatic lions. *The Indian Forester*, 121: 871.

Smithers, R. H. N. 1971. *The Mammals of Botswana*. The Trustees of the National Museums of Rhodesia, Salisbury.

Smithers, R. H. N. 1983. *The Mammals of the Southern African subregion*. University of Pretoria, Pretoria.

Smuts, G. L. 1976. Population characteristics and recent history of lions in two parts of the Kruger National Park. *Koedoe*, 19: 153–164.

Stander, P. E. 1992. Cooperative hunting in lions: the role of the individual. *Behavioral Ecology and Sociobiology*, 29: 445.

Stander, P. E., and S. D. Albon. 1993. Hunting success of lions in a semi-arid environment. In *Mammals as Predators*, edited by N. Dunstone and M. L. Gorman, 127–143. Zoological Society of London, London.

Turner, A., and M. Antón. 1997. *The Big Cats and Their Fossil Relatives: An Illustrated Guide to Their Evolution and Natural History*. Columbia University Press, New York.

Van Orsdol, K. G., J. P. Hanby, and J. D. Bygott. 1985. Ecological correlates of lion social organization (*Panthera leo*). *Journal of Zoology, London*, 206: 97–112.

Verschuren, J. 1958. *Écologie et biologie des grands mammifères (Primates, Carnivores, Ongulés)*. Institut des Parcs Nationaux du Congo Belge, Brussels.

Viljoen, P. C. 1993. The effects of changes in prey availability on lion predation in a large natural ecosystem in northern Botswana. In *Mammals as Predators*, edited by N. Dunstone and M. L. Gorman, 193–213. Zoological Society of London, London.

von Buol, P. 2000. "Buffalo lions": a feline missing link? *Swara*, 23: 20–25.

Wilson, V. J. 1975. *Mammals of the Wankie National Park, Rhodesia*. Trustees of the National Museums and Monuments of Rhodesia, Salisbury.

CHAPTER 7: The Lion's Mane: Geographic and Individual Variation

Anonymous. 2001. Löwenmänner ohne Mähne Mehrjährige Studien. *Geo*, 12: 216.

Astley Maberly, C. T. 1963. *The Game Animals of Southern Africa*. Thomas Nelson and Sons, Johannesburg.

Bath, M. O., and J. Chipperfield. 1969. *The Lions of Longleat*. Cassells & Company, London.

Blanford, W. T. 1870. *Observations on the Geology and Zoology of Abyssinia, Made during the Progress of the British Expedition to That Country in 1867–68*. Macmillan and Co., London.

Coheleach, G. 1982. *The Big Cats: The Paintings of Guy Coheleach*. Harry N. Abrams, Inc., New York.

Darwin, C. 1871. *The Descent of Man, and Selection in Relation to Sex*. D. Appleton and Company, New York.

Denis, A. 1964. *Cats of Africa: The African Lion*. Houghton Mifflin Company, Boston.

Ewer, R. F. 1973. *The Carnivores.* Comstock Publishing Associates, Ithaca, New York.

Grayson, D. K. 1990. Donner Party deaths: A demographic assessment. *Journal of Anthropological Research*, 46: 223–242.

Guggisberg, C. A. W. 1961. *Simba, the Life of the Lion*. Howard Timmins, Cape Town, South Africa.

Guggisberg, C. A. W. 1975. *Wild Cats of the World*. Taplinger Publishing Company, New York.

Guthrie, R. D. 1990. *Frozen Fauna of the Mammoth Steppe: The Story of Blue Babe.* University of Chicago Press, Chicago.

Haagner, A. K. 1920. *South African Mammals: A short manual for the use of field naturalists, sportsmen and travellers.* H. F. & G. Witherby, Cape Town, South Africa.

Harper, F. 1945. *Extinct and Vanishing Animals of the Old World*. American Committee for International Wildlife Protection, New York Zoological Park, New York.

Hemmer, H. 1962. Einiges über die Entstehung der Mähne des Löwen (*Panthera leo*) . *Säugetierkundliche Mitteilungen*, 10: 109–111.

Hemmer, H. 1974. Zur Artgeschichte des Löwen *Panthera (Panthera) leo* (Linnaeus, 1758) . *Veroff. Zool. Staatssamml. Munchen*, 17: 167–280.

Hollister, N. 1917. Some effects of environment and habit on captive lions. *Proceedings of the United States National Museum*, 53: 177–193.

Hunter, J. A. 1952. *Hunter.* Harper & Brothers, Publishers, New York.

Jefferson, G. T. 1992. The M_1 in *Panthera leo atrox*, an indicator of sexual dimorphism and ontogenetic age. *Current Research in the Pleistocene*, 9: 102–105.

Johnson, L. K. 1982. Sexual selection in a brentid weevil. *Evolution*, 36: 251–262.

Kingdon, J. 1977. *East African Mammals: An Atlas of Evolution in Africa*. Academic Press, London.

Kodric-Brown, A., and J. H. Brown. 1984. Truth in advertising: The kinds of traits favored by sexual selection. *The American Naturalist*, 124: 309–323.

Lönnberg, E. 1912. *Mammals Collected by the Swedish Zoological Expedition to British East Africa 1911*. Almquist & Wiksells, Uppsala & Stockholm.

MacDonald, D. W. 1984. The encyclopedia of mammals. in *The Encyclopedia of Mammals*, New York.

Mazák, V. 1964. Note on the lion's mane. *Zeitschrift für Säugetierkunde*, 29: 124–127.

Mazák, V., and A. M. Husson. 1960. Einige Bemerkungen über den Kaplöwen, *Panthera leo melanochaitus* (Ch. H. Smith, 1842) . *Zoologische Mededelingen (Leiden)*, 37: 101–111.

Packer, C., and J. Clottes. 2000. When lions ruled France. *Natural History*, 109: 52–57.

Peel, C. V. A. 1900. *Somaliland, Being an Account of Two Expeditions into the Far Interior.* F. E. Robinson & Co., London.

Pocock, R. I. 1931. The lions of Asia. *Journal of the Bombay Natural History Society*, 34: 638–665.

Roberts, A. F. 1986. *Animals in African Art.* The Museum for African Art, New York.

Roosevelt, T., and E. Heller. 1914. *Life-Histories of African Game Animals.* Charles Scribner's Sons, New York.

Selous, F. C. 1881. *A Hunter's Wanderings in Africa.* Richard Bentley & Son, London.

Selous, F. C. 1908. *African Nature Notes and Reminiscences.* Macmillan and Co., London.

Smithers, R. H. N. 1971. *The Mammals of Botswana.* The Trustees of the National Museums of Rhodesia, Salisbury.

Smithers, R. H. N. 1983. *The Mammals of the Southern African Subregion.* University of Pretoria, Pretoria.

Smuts, G. L., J. L. Anderson, and J. C. Austin. 1978. Age determination of the African lion (*Panthera leo*). *Journal of Zoology, London,* 185: 115–146.

Sommer, F. 1954. *Man and Beast in Africa.* Citadel Press, New York.

Stevenson-Hamilton, J. 1912. *Animal Life in Africa.* E. P. Dutton, New York.

Stevenson-Hamilton, J. 1947. *Wildlife in South Africa.* Cassel & Co., London.

Sutcliffe, A. J. 1985. *On the Track of Ice Age Mammals.* Harvard University Press, Cambridge, Massachusetts.

Swayne, H. C. G. 1895. *Seventeen Trips through Somaliland: A Tecord of Exploration & Big Game Shooting, 1885 to 1893.* Rowland Ward and Co., London.

Van Orsdol, K. G., J. P. Hanby, and J. D. Bygott. 1985. Ecological correlates of lion social organization (*Panthera leo*). *Journal of Zoology, London,* 206: 97–112.

Vereshchagin, N. K., and G. F. Baryshnikov. 1992. The ecological structure of the "Mammoth Fauna" in Eurasia. *Annales Zoologici Fennici,* 28: 253–259.

Vernay, A. S. 1930. The lion of India. *Natural History*: 30.

von Wolffe, J. F. 1955. *Mammals of Ethiopia and Principal Reptiles.* Rhodesian Litho, Salisbury, Southern Rhodesia.

West, P. M., and C. Packer. 2002. Sexual selection, temperature, and the lion's mane. *Science,* 297: 1339–1343.

Zahavi, A. 1975. Mate selection—a selection for a handicap. *Journal of Theoretical Biology,* 53: 205–214.

Zahavi, A. 1977. The cost of honesty (further remarks on the handicap principle). *Journal of Theoretical Biology,* 67: 603–605.

CHAPTER 8: Why the Lions of Tsavo Are Maneless

Astley Maberly, C. T. 1963. *The Game Animals of Southern Africa.* Thomas Nelson and Sons, Johannesburg.

Brown, J. L., M. Bush, and C. Packer. 1991. Developmental changes in pituitary-gonadal function in free-ranging lions (*Panthera leo leo*) of the Serengeti Plains and Ngorongoro Crater. *Journal of Reproduction and Fertility*, 91: 29.

Caputo, P. 2002. Maneless in Tsavo. *National Geographic Magazine*, 201: 38–53.

Chellam, R. 1996. Lions of the Gir Forest. *Wildlife Conservation*, 99: 40.

Clutton-Brock, T. H., F. E. Guinness, and S. D. Albon. 1982. *Red Deer.* University of Chicago Press, Chicago.

Dubach, J. M., B. D. Patterson, M. B. Briggs, K. Venzke, J. Flammand, P. Stander, L. Scheepers, and R. W. Kays. Molecular genetic variation across the southern geographic range of the African lion, *Panthera leo*. *Animal Conservation*, in review.

Dyer, A. 1973. *Classic African Animals: The Big Five.* Winchester Press, New York.

Ebling, F. J. 1987. The biology of hair. *Dermatologic Clinics*, 5: 467–481.

Eloff, F. C. 1973. Ecology and behavior of the Kalahari lion. In *The World's Cats*, edited by R. L. Eaton, 90–126. World Wildlife Safari, Winston, OR.

Gnoske, T., and J. Kerbis Peterhans. 2000. Cave lions: The truth behind biblical myths. *In the Field*, 71: 2–6.

Goss, R. J. 1983. Antlers: Regeneration, Function, and Evolution. Academic Press, New York.

Grinnell, J., C. Packer, and A. E. Pusey. 1995. Cooperation in male lions: Kinship, reciprocity or mutualism? *Animal Behaviour*, 49: 95.

Guggisberg, C. A. W. 1961. *Simba, the Life of the Lion.* Howard Timmins, Capetown, South Africa.

Guggisberg, C. A. W. 1975. *Wild Cats of the World.* Taplinger Publishing Company, New York.

Guthrie, R. D. 1990. *Frozen Fauna of the Mammoth Steppe: The Story of Blue Babe.* University of Chicago Press, Chicago.

Heinsohn, R., and C. Packer. 1995. Complex cooperative strategies in group-territorial African lions. *Science*, 269: 1260–1262.

Hubert, W. A. 1990. Psychotropic effects of testosterone. In *Testosterone: Action, Deficiency, Substitution*, edited by E. Nieschlag and H. M. Behre, 51–71. Springer-Verlag, New York.

Hunter, J. A. 1952. *Hunter.* Harper & Brothers, Publishers, New York.

Hunter, L. 1999. No mane event. *BBC Wildlife*: 26.

Johnson, M. 1928. *Safari: A Saga of the African Blue.* G. P. Putnam's Sons, New York.

Kays, R. W., and B. D. Patterson. 2002. Mane variation in African lions and its social correlates. *Canadian Journal of Zoology*, 80: 471–478.

Kodric-Brown, A., and J. H. Brown. 1984. Truth in advertising: The kinds of traits favored by sexual selection. *The American Naturalist*, 124: 309–323.

Leuthold, W., and B. M. Leuthold. 1976. Density and biomass of ungulates in Tsavo East National Park. *East African Wildlife Journal*, 14: 49–58.

Leyhausen, P. 1995. A further note on maneless lions. *Social Biology and Human Affairs*, 60: 56.

Lönnberg, E. 1912. *Mammals Collected by the Swedish Zoological Expedition to British East Africa 1911*. Almquist & Wiksells, Uppsala & Stockholm.

Lorenz, K. 1966. *On Aggression*. Harcourt Brace & World, New York.

Lumia, A. R., K. M. Thorner, and M. Y. Mcginnis. 1994. Effects of chronically high doses of the anabolic androgenic steroid, testosterone, on intermale aggression and sexual behavior in male rats. *Phys. Behav.*, 55: 331–335.

Mazák, V. 1970. The Barbary lion, *Panthera leo leo* (Linnaeus, 1758); some systematic notes, and an interim list of the specimens preserved in European museums. *Zeitschrift für Säugetierkunde*, 35: 34–44.

McBride, C. 1977. *The White Lions of Timbavati*. Paddington Press, New York, NY.

Millar, K. V., R. L. Marchinton, K. J. Forand, and K. L. Johansen. 1987. Dominance, testosterone levels, and scraping activity in captive herd of white-tailed deer. *Journal of Mammalogy*, 68: 812–817.

Munson, L., J. L. Brown, and D. E. Wildt. 1996. Genetic diversity affects testicular morphology in free-ranging lions (*Panthera leo*) of the Serengeti Plains and Ngorongoro Crater. *Journal of Reproduction and Fertility*, 108: 11.

Packer, C., D. Scheel, and A. E. Pusey. 1990. Why lions form groups: food is not enough. *The American Naturalist*, 136: 1–19.

Patterson, B. D., J. M. Dubach, T. P. Gnoske, S. Weru, E. Mwangi, and R. W. Kays. Year. *Morphologic, Genetic, and Ecological Variation of African Lions: The Mane Story*. 79th Annual Meeting of the American Society of Mammalogists, University of Washington, Seattle, WA, 17–21 June 1999.

Pocock, R. I. 1931. The lions of Asia. *Journal of the Bombay Natural History Society*, 34: 638–665.

Rachlow, J. L., E. V. Berkeley, and J. Berger. 1998. Correlates of male mating strategies in white rhinos (*Ceratotherium simum*). *Journal of Mammalogy*, 79: 1317–1324.

Roosevelt, T., and E. Heller. 1914. *Life-Histories of African Game Animals.* Charles Scribner's Sons, New York.

Russell, W. M. S. 1994. Greek and Roman monsters. Part 2. *Social Biology and Human Affairs*, 59: 1–8.

Sauer, P. 1984. Physical characteristics. In *White-tailed Deer: Ecology and Management*, 73–90. Stackpole Books, Harrisburg, PA.

Schaller, G. B. 1972. *The Serengeti Lion: A Study of Predator-Prey Relations.* University of Chicago Press, Chicago.

Selous, F. C. 1881. *A Hunter's Wanderings in Africa*. Richard Bentley & Son, London.

Smuts, G. L., J. L. Anderson, and J. C. Austin. 1978. Age determination of the African lion (*Panthera leo*). *Journal of Zoology, London*, 185: 115–146.

Sommer, F. 1954. *Man and Beast in Africa*. Citadel Press, New York.

Stevenson-Hamilton, J. 1912. *Animal Life in Africa*. E. P. Dutton, New York.

Taylor, J. 1959. *Maneaters and Marauders.* A. S. Barnes and Co., New York.

Thornton, M. J., I. Laing, K. Hamada, A. G. Messenger, and V. A. Randall. 1993. Differences in testosterone metabolism by beard and scalp hair follicle dermal papilla cells. *Clinical Endocrinology*, 39: 633–639.

Van Orsdol, K. G. 1984. Foraging behaviour and hunting success of lions in Queen Elizabeth National Park, Uganda. *African Journal of Zoology*, 22: 79–99.

van Wijngaarden, W., and V. W. P. van Engelen. 1989. *Soils and vegetation of the Tsavo area*. Reconnaissance Soil Survey, Kenya Soil Survey series no. 7, Kenya Soil Survey, Nairobi.

Vaughan, T. A., J. M. Ryan, and N. J. Czaplewski. 2000. *Mammalogy*. 4th ed. Saunders College Publishing, Philadelphia.

von Buol, P. 2000. "Buffalo lions": A feline missing link? *Swara*, 23: 20–25.

West, P. M., and C. Packer. 2002. Sexual selection, temperature, and the lion's mane. *Science*, 297: 1339–1343.

Woodroffe, R., D. W. MacDonald, and C. L. Cheeseman. 1997. Endocrine correlates of contrasting male mating strategies in the European badger (*Meles meles*). *Journal of Zoology, London*, 241: 291–300.

CHAPTER 9: Conservation and Tsavo National Parks

African Lion Working Group. 2001. Hunting ban in Botswana. *African Lion News*, 3: 1.

Andanje, S. A., and W. K. Ottichilo. 1999. Population status and feeding habits of the translocated sub-population of Hunter's antelope or hirola (*Beatragus hunteri* [Sclater, 1889]) in Tsavo East National Park, Kenya. *African Journal of Ecology*, 37: 38–48.

Anonymous. 1997. Man-eating lions shot in Kruger National Park. *Cat News*, 27: 11.

Anonymous. 1999. Zimbabwe apologises over Briton mauled by lions. In *Zimbabwe apologises over Briton mauled by lions*. Johannesburg, South Africa.

Beard, P. H. 1988. *The End of the Game: The Last Word from Paradise*. Thames & Hudson Ltd., London.

Belsky, A. J. 1992. Effects of grazing, competition, disturbance, and fire on species composition and diversity in grassland communities. *Journal of Vegetation Science*, 3: 187–200.

Brooks, T. M., S. L. Pimm, and J. O. Oyugi. 1999. Time lags between deforestation and bird extinction in tropical forest fragments. *Conservation Biology*, 13: 1140–1150.

Caro, T. 2000. Focal species. *Conservation Biology*, 14: 1569–1570.

Durrheim, D. N., and P. A. Leggat. 1999. Risk to tourists posed by wild mammals in South Africa. *Journal of Travel Medicine*, 6: 172–179.

Eloff, F. C. 1973. Ecology and behavior of the Kalahari lion. In *The World's Cats*, edited by R. L. Eaton, 90–126. World Wildlife Safari, Winston, OR.

Frank, L. G. 1998. Living with lions: Carnivore conservation and livestock in Laikipia District, Kenya. U.S. Agency for International Development, Conservation of Biodiverse Resource Areas Project, 623-0247-C-00-3002-00, Mpala Research Centre, Nanyuki, Kenya.

Graham, A., and P. H. Beard. 1973. *Eyelids of Morning: The Mingled Destinies of Crocodiles and Men.* Chronicle Books, San Francisco.

Homer. 1966. *The Iliad,* translated by E. V. Rieu. Penguin Books, Baltimore, MD.

Hunter, J. A. 1952. *Hunter.* Harper & Brothers, Publishers, New York.

Jackson, P. 1997. Status of the tiger *Panthera tigris* (Linnaeus 1758) in 1997. *Cat News,* 27: 10.

Kahumbu, P., P. Omond, I. Douglas-Hamilton, and J. King. 1999. Total aerial count of elephants in the Tsavo National Park and adjacent areas, January 1999 report, Kenya Wildlife Service, Nairobi.

Karani, I. W. 1994. An assessment of depredation by lions and other predators in the group ranches adjacent to Masai Mara National Reserve. Unpublished thesis, Department of Wildlife Management,Moi University, Eldoret (Kenya).

Laws, R. M. 1974. Behaviour, dynamics and management of elephant populations. In *The Behaviour of Ungulates and Its Relations to Management,* edited by V. Geist and F. Walther, 513–529. International Union for Conservation of Nature and Natural Resources 24, Morges, Switzerland

Lens, L., S. Van Dongen, C. M. Wilder, T. M. Brooks, and E. Matthysen. 1999. Fluctuating asymmetry increases with habitat disturbance in seven bird species of a fragmented afrotropical forest. *Proceedings of the Royal Society of London, Series B,* 266: 1241–1246.

Macleod, F. 2001. Bush guns for Botswana lion hunt. In *Bush guns for Botswana lion hunt,* Johannesburg.

Macleod, F. 2002. From Canned to "Candy-coated" Hunting. In *From Canned to 'Candy-coated' Hunting,* Johannesburg.

McNaughton, S. J., and N. J. Georgiadis. 1986. Ecology of African grazing and browsing mammals. *Annual Review of Ecology and Systematics,* 17: 39–65.

Myers, N., R. A. Mittermeier, C. G. Mittermeier, G. A. B. da Fonseca, and J. Kent. 2000. Biodiversity hotspots for conservation priorities. *Nature,* 403: 853–858.

Newmark, W. D. 1995. Extinction of mammal populations in western North American national parks. *Conservation Biology,* 9: 512–526.

Newmark, W. D., D. N. Manyanza, D.-G. M. Gamassa, and H. I. Sariko. 1994. The conflict between wildlife and local people living adjacent to protected areas in Tanzania: Human density as a predictor. *Conservation Biology,* 8: 249–255.

Patterson, B. D., S. M. Kasiki, E. Selempo, and R. W. Kays. In press. Livestock predation by lions (*Panthera leo*) on ranches neighboring Tsavo East National Park, Kenya. Biological Conservation.

Patterson, J. H. 1907. *The Man-Eaters of Tsavo and Other East African Adventures.* Macmillan, London.

Quammen, D. 1996. *The Song of the Dodo: Island Biogeography in the Age of Extinctions.* Scribner, New York.

Roosevelt, T. 1909. *African Game Trails. An Account of the African Wanderings of an American Hunter-Naturalist.* Charles Scribner's Sons, New York.

Selous, F. C. 1881. *A Hunter's Wanderings in Africa.* Richard Bentley & Son, London.

Sheldrick, D. 2000. Vegetation changes in Tsavo National Park, Kenya, 1885–1996: Elephant densities and management. *Elephant*, 2: 26–33.

Simon, N. 1962. *Between the Sunlight and the Thunder: The Wild Life of Kenya.* Collins, London.

Stander, P. E. 1990. A suggested management strategy for stock-raiding lions in Namibia. *South African Journal of Wildlife Research*, 20: 53–60.

Western, D. 1997. *In the Dust of Kilimanjaro.* Island Press/Shearwater Books, Washington, DC.

Yamazaki, K. 1996. Social variation of lions in a male-depopulated area in Zambia. *Journal of Wildlife Management*, 60: 490–497.

Acknowledgments

Caputo, P. 2002. Ghosts of Tsavo: *Stalking the Mystery Lions of East Africa.* Washington, D.C., Adventure Press National Geographic.

Kerbis Peterhans, J. C., C. M. Kusimba, T. P. Gnoske, S. A. Andanje, and B. D. Patterson. 1998. Man-eaters of Tsavo. *Natural History*, 107: 12-14.

Kerbis Peterhans, J. C., and T. P. Gnoske. 2002. The science of "man-eating" among lions *Panthera leo* with a reconstruction of the natural history of the 'man-eaters of Tsavo'. *Journal of East African Natural History* 90 ["2001"]:1-40.

Kusimba, C. M., J. C. Kerbis Peterhans, T. P. Gnoske, and B. D. Patterson, "Excavations of the 'Man-eaters' Cave', Tsavo, Kenya," Eli Lilly Foundation, Indianapolis, IN, 1998.

Patterson, B. D., C. M. Kusimba, J. C. Kerbis Peterhans, T. P. Gnoske, and D. Brinkmeier, "The Tsavo Initiative: linking people and wildlife in a dynamic environment," Proposal to Kenya Wildlife Service, 1998.

Index